HANNU VÄLIAHO · TIMO PEKKONEN

A Procedure for Stepwise Regression Analysis

A Procedure for Stepwise Regression Analysis

(with a Program in FORTRAN V)

by

Dr. HANNU VÄLIAHO
and TIMO PEKKONEN

Forest Research Institute Helsinki, Finnland

AKADEMIE-VERLAG · BERLIN
1976

Erschienen im Akademie-Verlag, 108 Berlin, Leipziger Str. 3—4

Lizenznummer: 202 · 100/424/76
Gesamtherstellung: VEB Druckerei „Thomas Müntzer", 582 Bad Langensalza
Bestellnummer: 7622122 (6303) · LSV 1075
Printed in GDR
EVP 18,–

Preface

The aim of this book is to introduce a method for stepwise regression analysis developed by VÄLIAHO [4], and give a computer program for this method and an additional program for data transformation.

The method is stepwise with respect to both observations and variables: one observation or independent variable at a time can be added to the model or removed from it. Restricted regression analysis will also be considered: exact and approximate linear restrictions between the regression coefficients can be added or removed one at a time. Observations, independent variables, and exact and approximate restrictions may be added and removed sequentially in any order.

The computer program for stepwise regression analysis is written in FORTRAN V for time sharing. Weighting, semi-automatic variable selection and discounting of observations are included in the program. The transformation program consists of a ready coded main program and a subprogram to be coded separately.

Of the authors, H. VÄLIAHO has written the theoretical part of the book (Chaps. 1—4). The basic ideas underlying the computer programs (Chaps. 5—6) are also his, while T. PEKKONEN has carried through the systems design and programming of these programs.

HANNU VÄLIAHO TIMO PEKKONEN

Contents

In regression analysis it is sometimes useful to add or remove observations one at a time, the set of independent variables being fixed. In some connections, on the other hand, it is desirable to add or remove independent variables one at a time, the set of observations being fixed. The stepwise method to be introduced is a synthesis of these two approaches.

In the stepwise method, processing an observation means changing its weight in the model. This includes adding and removing an observation as particular cases. The observation processed in a step together with the change of its weight constitutes a data-piece. Processing an observation can also be expressed by saying that the corresponding data-piece is added to the model. Processing a variable means shifting it from the set of dependent variables to the set of independent variables or vice versa. Processing of observations and variables can be performed in any order.

In the stepwise method, certain transformations are applied to the so-called model scheme, a matrix containing one row and column for each variable, independent or dependent. At a given stage the scheme consists of the inverse of the moment matrix of the independent variables, the regression coefficients vector for each dependent variable and the residual moment matrix. As a rule, the recursion formulae are derived for multiple stepping; formulae for single stepping are obtained as particular cases. In addition to the model scheme, an extended scheme is introduced, including, in addition to the model scheme, one row and column for each data-piece. In this scheme, observations and variables may be processed by means of the same transformation, viz. the pivotal operation. Thus the pivotal operation constitutes a corner-stone of the stepwise method.

In the book, restricted regression analysis is also considered. Restrictions express a priori knowledge about the regression coefficients in the form of exact or approximate linear equations, called equality restrictions and desiderata, respectively. The closeness of the desiderata equations is regulated by weights associated with them. If these weights tend to infinity, the desiderata are changed into equality restrictions. When estimating parameters of the regression model the desiderata are taken into account in the same way as observations. Formulae for adding equality restrictions are obtained as limiting cases from the corresponding formulae for desiderata, letting their weights tend to infinity. In removing equality restrictions, the extended scheme is needed; it suffices to use a modified form, containing additional rows and columns for equality restrictions only. In this scheme, equality restrictions may be added and removed by pivotal operations. In restricted regression analysis observations, variables, desiderata and equality restrictions may be processed sequentially in any order.

A computer program for the stepwise method is included in the book, as well as a program for data transformation. The former is time-sharing oriented, which makes semi-automatic variable selection possible. Weighting and discounting observations are also included. As input material, an observation matrix and/or a moment matrix may be used. The modular structure allows for flexible addition of additional features to the program. The transformation program prepares input material (both transformed observation matrices and moment matrices) to the regression program. It consists of a ready coded main program and a subprogram to be coded separately (the desired transformations are included in the subprogram).

1. Mathematical Background[1]

1.1. Sets and matrices

1° Sets. Let $I = \{i_1, i_2, \dots, i_n\}$ be a set of integers. If an element i belongs to I, $i \in I$ is written. If I_1 is a subset of I, $I_1 \subset I$ is written. The integers $i \in I$ may refer to numbered objects, such as the rows and columns of a matrix.

The union of sets I and J is denoted by $I \cup J$, and the intersection by $I \cap J$. If $J \subset I$, set $\bar{J} = I - J$ is termed the complement of J with respect to I. The empty set is denoted by $\emptyset$. If $I \cap J = \emptyset$, set $I \cup J$ may also be denoted by $I + J$.

2° Matrix notation. A matrix of order $m \times n$ is denoted by $A = A_{m,n} = [a_{ij}]$. In A, row i is denoted by A_i and column j by a_j. A column vector, i.e., an $m \times 1$ matrix, is denoted by $x = x_{m,1}$. A matrix $A_{n,n}$ is a square matrix of order n. For a diagonal matrix $D_{n,n}$, the abbreviated notation $D = \lfloor \overline{d_1 \cdots d_n} \rfloor$ will be used. An identity matrix is denoted by I, and a zero matrix by 0. A prime indicates transposition. A square matrix A is called symmetric if $A' = A$. The inverse of a square matrix A is denoted by A^{-1} (if existent).

Let $A = A_{m,n}$, $I \subset \{1, \dots, m\}$ and $J \subset \{1, \dots, n\}$. A_{IJ} is used to denote the submatrix obtained from A by deletion of the rows $i \in \{1, \dots, m\} - I$, and the columns $j \in \{1, \dots, n\} - J$. Furthermore, we denote $A_I = A_{I\{1,\dots,n\}}$ and $x_I = x_{I\{1\}}$. If A is square, a submatrix of the form A_{II} is called a principal submatrix of A.

3° Partitioned matrices. Let a matrix A be partitioned as follows

$$A = [A_{ij}] = \begin{bmatrix} A_{11} & \cdots & A_{1n} \\ \vdots & & \vdots \\ A_{m1} & \cdots & A_{mn} \end{bmatrix}.$$

There are m super-rows, and n super-columns. On occasion, blocks A_{ij} are known as super-elements. If the super-row i, $i = 1, \dots, m$, consists

[1] A more comprehensive mathematical introduction is given in Väliaho [4]. Reference is also made to Wendler [7].

of p_i rows, and the super-column j, $j = 1, \dots, n$, of q_j columns, it is said that A is of the order $(p_1, p_2, \dots, p_m) \times (q_1, q_2, \dots, q_n)$. If A is square, and $m = n$ and $p_i = q_i$ for all $i = 1, \dots, m$, it may be said that A is of the order $(p_1, p_2, \dots, p_m)$.

1.2. The pivotal operation

Let us consider the transformation

$$- y = Ax \,, \tag{1}$$

where A is $n \times n$, and x and y are $n \times 1$. On partitioning, we write

$$\begin{bmatrix} -y_1 \\ -y_2 \end{bmatrix} = \begin{bmatrix} A_{11} & A_{12} \\ A_{21} & A_{22} \end{bmatrix} \begin{bmatrix} x_1 \\ x_2 \end{bmatrix}, \tag{2}$$

where A_{11} is $k \times k$, or, in other words,

$$\begin{aligned} -y_1 &= A_{11}x_1 + A_{12}x_2 \,, \\ -y_2 &= A_{21}x_1 + A_{22}x_2 \,. \end{aligned} \tag{3}$$

Assume that A_{11} is non-singular; then x_1 may be solved from (3a), and substituted into (3b):

$$- x_1 = A_{11}^{-1}y_1 + A_{11}^{-1}A_{12}x_2 \,,$$

$$- y_2 = - A_{21}A_{11}^{-1}y_1 + (A_{22} - A_{21}A_{11}^{-1}A_{12})\, x_2$$

or equivalently

$$\begin{bmatrix} -x_1 \\ -y_2 \end{bmatrix} = \begin{bmatrix} A_{11}^{-1} & A_{11}^{-1}A_{12} \\ -A_{21}A_{11}^{-1} & A_{22} - A_{21}A_{11}^{-1}A_{12} \end{bmatrix} \begin{bmatrix} y_1 \\ x_2 \end{bmatrix}.$$

It is noted that x_1 and y_1 have been interchanged in comparison with (2). At the same time, the matrix of the transformation has changed as follows:

$$\begin{bmatrix} A_{11} & A_{12} \\ A_{21} & A_{22} \end{bmatrix} \to \begin{bmatrix} A_{11}^{-1} & A_{11}^{-1}A_{12} \\ -A_{21}A_{11}^{-1} & A_{22} - A_{21}A_{11}^{-1}A_{12} \end{bmatrix}.$$

The operation during which a transformation of this kind is accomplished will be termed a pivotal operation. We give:

Definition 1.2.1: The standardized pivotal operation $\mathcal{P}_{\{1,\dots,k\}}$ on any n-square $(0 < k < n)$ matrix A is defined by

$$\mathcal{P}_{\{1,\dots,k\}}A = \mathcal{P}_{\{1,\dots,k\}} \begin{bmatrix} A_{11} & A_{12} \\ A_{21} & A_{22} \end{bmatrix} = \begin{bmatrix} A_{11}^{-1} & A_{11}^{-1}A_{12} \\ -A_{21}A_{11}^{-1} & A_{22} - A_{21}A_{11}^{-1}A_{12} \end{bmatrix},$$

where A_{11} (to be called the pivot matrix) is $k \times k$, provided that A_{11}^{-1} exists. Furthermore, we define

$$\mathcal{P}_{\emptyset}A = A \quad \text{and} \quad \mathcal{P}_{\{1,\dots,n\}}A = A^{-1} \text{ (if existent)} .$$

In the case of $\mathcal{P}_{\emptyset}$, no exchange of the variables x_i, y_i occurs, and in the case of $\mathcal{P}_{\{1,\dots,n\}}$, the whole vectors x and y are interchanged.

The general pivotal operation is defined as follows:

Definition 1.2.2: The pivotal operation $\mathcal{P}_K$, $K \subset \{1, \dots, n\}$, is defined as the operation during which the variables x_K and y_K in (1) are interchanged.

Remark 1.2.1: Let $K = \{i_1, i_2, \dots, i_k\}$. For the performance of a pivotal operation, the variables may be renumbered so that x_{i_h}, y_{i_h} are termed, say, x'_h, y'_h; $\mathcal{P}_{\{1,\dots,k\}}$ is then performed, and the original numbering is finally returned.

Let then $A = [A_{ij}]$ be a partitioned matrix, with the submatrices A_{ii} principal. Then $\mathcal{P}_{[k]}$, the pivotal operation with A_{kk} as the pivot matrix, takes the following form:

$$\mathcal{P}_{[k]}A = A^+ \text{ with } \begin{cases} A_{kk}^+ = A_{kk}^{-1} , & \\ A_{kj}^+ = A_{kk}^{-1}A_{kj} , & j \neq k , \\ A_{ik}^+ = -A_{ik}A_{kk}^{-1} , & i \neq k , \\ A_{ij}^+ = A_{ij} - A_{ik}A_{kk}^{-1}A_{kj} , & i, j \neq k . \end{cases}$$

Another expression for "applying $\mathcal{P}_{[k]}$" is "pivoting at A_{kk}". If K is reduced to only one element k, a single pivotal operation ensues:

$$\mathcal{P}_kA = A^+ \text{ with } \begin{cases} a_{kk}^+ = 1/a_{kk} , & \\ a_{kj}^+ = a_{kj}/a_{kk} , & j \neq k , \\ a_{ik}^+ = -a_{ik}/a_{kk} , & i \neq k , \\ a_{ij}^+ = a_{ij} - a_{ik}a_{kj}/a_{kk} , & i, j \neq k . \end{cases} \tag{4}$$

Here a_{kk} is assumed to be non-zero.

In the following theorems, the tacit assumption is made that the pivotal operations appearing are defined.

Theorem 1.2.1: *Any pivotal operation $\mathcal{P}$ is involutory, i.e., $\mathcal{P}^2A = A$.*

P r o o f: x_K and y_K are first interchanged, and then returned to their original places.

Definition 1.2.3: Along with operation $\mathcal{P}_K$, we define the indicator matrix

$$\Psi = \overline{\underline{|\psi_1 \cdots \psi_n|}} \quad \text{with } \psi_i = \begin{cases} -1 & \text{if } i \in K , \\ 1 & \text{otherwise} . \end{cases}$$

Denoting $\mathcal{P}_\Psi = \mathcal{P}_K$, we have

Theorem 1.2.2: *For any two pivotal operations $\mathcal{P}_{\Psi_1}$ and $\mathcal{P}_{\Psi_2}$, we have*

$\mathcal{P}_{\Psi_2}\mathcal{P}_{\Psi_1} = \mathcal{P}_{\Psi_1}\mathcal{P}_{\Psi_2} = \mathcal{P}_{\Psi_1\Psi_2}$.

Proof: Let first $K_1 \cap K_2 = \emptyset$. It is apparent that the following exchanges give the same result:

(i) $x_{K_1} \leftrightarrow y_{K_1}$ and $x_{K_2} \leftrightarrow y_{K_2}$,

(ii) $x_{K_2} \leftrightarrow y_{K_2}$ and $x_{K_1} \leftrightarrow y_{K_1}$,

(iii) $x_{K_1+K_2} \leftrightarrow y_{K_1+K_2}$.

Let then $K_1 \cap K_2 = K_3 \neq \emptyset$. In this case, $\mathcal{P}_{\Psi_2}\mathcal{P}_{\Psi_1}$ may be decomposed into the following exchanges:

$x_{K_1-K_3} \leftrightarrow y_{K_1-K_3}$; $x_{K_3} \leftrightarrow y_{K_3}$; $y_{K_3} \leftrightarrow x_{K_3}$; $x_{K_2-K_3} \leftrightarrow y_{K_2-K_3}$.

The intermediate exchanges are cancelled, and thus the combined exchange equals

$x_{K_4} \leftrightarrow y_{K_4}$ with $K_4 = K_1 \cup K_2 - K_1 \cap K_2$

corresponding to $\mathcal{P}_{\Psi_1\Psi_2}$. However,

$\mathcal{P}_{\Psi_1}\mathcal{P}_{\Psi_2} = \mathcal{P}_{\Psi_1\Psi_2} = \mathcal{P}_{\Psi_2}\mathcal{P}_{\Psi_1}$

The preceding theorem may be generalized:

Theorem 1.2.3: *The product of pivotal operations is a pivotal operation of which the indicator matrix is the product of the indicator matrices of the original pivotal operations. In other words,*

$\mathcal{P}_{\Psi_n} \cdots \mathcal{P}_{\Psi_1} = \mathcal{P}_{\Psi_1 \cdots \Psi_n}$.

The operations on the left may be performed in any order.

2. Linear Regression Analysis

2.1. The theoretical regression model

Suppose that we have at our disposal an observational material

$M = [W, Z] = [W, X, Y]$,

where

$Z_{n,m}$	is the observation matrix,
$X_{n,k}$	is the matrix of observations on the independent variables,
$Y_{n,m-k}$	with the columns numbered $(k+1), \dots, m$, is the matrix of observations on the dependent variables,

$W = \lceil w_1 \cdots w_n \rfloor = \lceil v_1^2 \cdots v_n^2 \rfloor$ is the weighting matrix.

The rows Z_i stand for observations and the columns z_j for variables. Thus the columns $x_j = z_j$, $j = 1, \dots, k$, and $y_j = z_j$, $j = (k+1), \dots, m$, stand for independent and dependent variables, respectively.

On the material M we base the theoretical regression models

$$y_j = X\beta_j + \varepsilon_j \qquad j = (k+1), \dots, m \tag{1}$$

or, combined,

$$Y = X\beta + \varepsilon, \tag{2}$$

where

$$\beta = [\beta_{k+1} \cdots \beta_m] = \begin{bmatrix} \beta_{1,k+1} & \cdots & \beta_{1m} \\ \vdots & & \vdots \\ \beta_{k,k+1} & \cdots & \beta_{km} \end{bmatrix}$$

is the matrix of theoretical regression coefficients and

$$\varepsilon = [\varepsilon_{k+1} \cdots \varepsilon_m] = \begin{bmatrix} \varepsilon_{1,k+1} & \cdots & \varepsilon_{1m} \\ \vdots & & \vdots \\ \varepsilon_{n,k+1} & \cdots & \varepsilon_{nm} \end{bmatrix}$$

is the error matrix. Note that β_{ij} is the regression coefficient of y_j on x_i and ε_{ij} is the error of y_j in the i:th observation. The absence of a constant term in (1) does not limit the scope: we may, for example, take $x_{i1} = 1$ for all i, when β_{1j} is converted into a constant term. In connection with model (2) we say that the variables $x_1, \ldots, x_k$ are in regression while $y_{k+1}, \ldots, y_m$ are not in regression.

In general, only the linear independence of the columns of X is assumed. When needed, some of the following additional assumptions will be adopted:

(i) X (given) is non-stochastic,

(ii) Y (observed) is stochastic,

(iii) $W = V^2$ (given) is non-stochastic,

(iv) β (unknown) is non-stochastic,

(v) ε (unknown) is stochastic,

(vi) $\mathsf{E}(\varepsilon) = 0$,

(vii) $\mathsf{E}(V\varepsilon_p)\,(V\varepsilon_j)' = \begin{cases} 0 & \text{if } p \neq j, \\ \sigma_j^2 I & \text{if } p = j, \end{cases}$

(viii) $V\varepsilon_j$ is $N(0, \sigma_j^2 I)$, $\quad j = (k+1), \ldots, m$.

Above, σ_j^2 is termed the error variance, and assumption (viii) is called the normality assumption.

2.2. The empirical regression model

The empirical counterpart of (2.1.2) is the model

$$Y = XB + E\,, \tag{1}$$

where B is an estimate of β and E is the residual matrix. The regression coefficients are estimated by the weighted least squares method, i.e., the matrix B is determined so that for $j = (k+1), \ldots, m$

$$\bar{q}_{jj} = (Ve_j)'\,(Ve_j) = (y_j - X\bar{b}_j)'\,W(y_j - X\bar{b}_j) \tag{2}$$

is minimum with respect to $\bar{b}_j$. This yields

$$\begin{aligned} B &= (X'WX)^{-1}\,X'WY = UX'WY\,, \\ E &= Y - XB = (I - XUX'W)\,Y\,, \\ Q &= E'WE = Y'(W - WXUX'W)\,Y\,, \end{aligned} \tag{3}$$

where

$$U = (X'WX)^{-1} = \begin{bmatrix} u_{11} & \cdots & u_{1k} \\ \vdots & & \vdots \\ u_{k1} & \cdots & u_{kk} \end{bmatrix}$$

is the inverse of the (weighted) moment matrix $X'WX$ of model (1), and

$$Q = \begin{bmatrix} q_{k+1,k+1} & \cdots & q_{k+1,m} \\ \vdots & & \vdots \\ q_{m,k+1} & \cdots & q_{mm} \end{bmatrix}$$

is the (weighted) residual moment matrix of the model.

2.3. Technical aspects

For computational reasons the information from model (2.2.1) is collected into the matrix of the model

$$A = \begin{bmatrix} U & B \\ -B' & Q \end{bmatrix} \tag{1}$$

also called briefly the model, where the rows and columns correspond to the variables $x_1, \dots, x_k, y_{k+1}, \dots, y_m$, in this order. Note that $A = \mathcal{P}_K Z'WZ$ with $K = \{1, \dots, k\}$.

Next we generalize the concept of the matrix of a model, cf. (1), letting any set $R \subset \{1, \dots, m\}$ of the variables be independent variables. Apparently we have $A = \mathcal{P}_R Z'WZ$. The indicator matrix Ψ corresponding to R may now be called also the indicator matrix of model A. We have

$$\Psi = \lceil \psi_1 \cdots \psi_m \rfloor \text{ with } \psi_j = \begin{cases} -1 & \text{if } z_j \text{ is in regression,} \\ 1 & \text{if } z_j \text{ is not in regression.} \end{cases} \tag{2}$$

Initially, $\Psi = I$. At any stage, the elements of A have the following meanings:

$$a_{ij} = \begin{cases} u_{ij} & \text{if } \psi_i = \psi_j = -1\,, \\ b_{ij} & \text{if } \psi_i = -1,\ \psi_j = 1\,, \\ -b_{ji} & \text{if } \psi_i = 1,\ \psi_j = -1\,, \\ q_{ij} & \text{if } \psi_i = \psi_j = 1\,. \end{cases} \tag{3}$$

2.4. Multiple and partial correlation

The goodness of fit of the model $y_j = Xb_j + e_j$, $j = (k+1), \dots, m$, is measured by means of the multiple correlation coefficient

$$R_j = \sqrt{1 - s_j^2/s_{0j}^2}\,, \tag{1}$$

where

$$s_j^2 = \frac{q_{jj}}{n-k} \tag{2}$$

to be called the residual variance, is an estimator of σ_j^2, and

$$s_{0j}^2 = \frac{\sum_{i=1}^{n} w_i(y_{ij} - \bar{y}_j)^2}{n-1}$$

with

$$\bar{y}_j = \frac{\sum_{i=1}^{n} w_i y_{ij}}{\sum_{i=1}^{n} w_i}$$

is the variance of y_j.

The partial correlation coefficient r_{ij} of z_i and z_j, adjusted for $x_1, \ldots, x_k$, is calculated from

$$r_{ij} = q_{ij}/\sqrt{q_{ii}q_{jj}} \,. \tag{3}$$

This statistic is easily calculated from the matrix of the model.

Finally, we note that in order to calculate multiple or partial correlation coefficients it is recommended that the constant term be included in the model.

2.5. Distributions of parameters

The matrices B, E, Q in (2.2.3) are obtained directly by means of the least squares method and are valid, regardless of the assumptions of Sec. 2.1. Only the linear independence of the columns of X is required.

Adopting the assumptions (i—vii), we find that B is an unbiased estimator of β, i.e. $\mathsf{E}(B) = \beta$, and $\mathsf{E}(E) = 0$. Moreover, (2.4.2) gives an unbiased estimator of σ_j^2. We also have

$$\text{cov}\,(b_j) = \mathsf{E}[(b_j - \beta_j)\,(b_j - \beta_j)'] = \sigma_j^2 U \qquad j = (k+1), \ldots, m\,. \tag{1}$$

An estimator for cov (b_j) is obtained from (1) by substituting s_j^2 for σ_j^2.

The additional adoption of the normality assumption produces for $i = 1, \dots, k;\ j = (k+1), \dots, m$

(i) the coefficient b_{ij} is $N(\beta_{ij}, \sigma_j^2 u_{ii})$,

(ii) the ratio $\dfrac{b_{ij} - \beta_{ij}}{s_j \sqrt{u_{ii}}}$ is distributed as t_{n-k},

(iii) the residual moment q_{jj} is distributed as $\sigma_j^2 \chi_{n-k}^2$.

In addition to the model A in (2.3.1), consider the model A^* obtained from A by removing the independent variables $x_1, \dots, x_p$. Then:

(iv) Under the hypothesis $\beta_{1j} = \dots = \beta_{pj} = 0$ the ratio

$$F = \frac{(q_{jj}^* - q_{jj})/p}{q_{jj}/(n-k)} \tag{2}$$

is distributed as FISHER's $F(p, n-k)$.

Note that using the result (iv) it is possible to test whether the removed independent variables were significant in model A.

Assumption (vii) of Sec. 2.1. indicates that weighting is used for stabilizing the variances of different observations. If a different weighting is applied, then the estimators of the variances of regression coefficients and the estimator of the residual variance are biased, and the results (i—iv) above are no more valid; see VÄLIAHO [5].

3. Restricted Regression Analysis

3.1. Principles of restricted models

A regression model is said to be restricted if the regression coefficients are subject to certain linear restrictions: desiderata, equality restrictions, or inequality restrictions. These restrictions express a priori knowledge about the regression coefficients. They are defined below in connection with a regression model $Y = X\beta + \varepsilon$ including k independent and $m - k$ dependent variables.

(i) Desiderata are approximate linear restrictions $R\beta + \Phi = C$, where R is $p \times k$, and $C = [c_{k+1} \cdots c_m]$ and $\Phi = [\varphi_{k+1} \cdots \varphi_m]$ are $p \times (m - k)$. Φ is a matrix of errors. For the desiderata a weighting matrix $\Gamma = \overline{|\gamma_1 \cdots \gamma_p|}$ is assumed to be given.

(ii) Equality restrictions, called simply restrictions if no danger of confusion exists, are exact linear restrictions $R\beta = C$.

(iii) Inequality restrictions are linear inequalities $R\beta_j \geqq c_j$.

The consept of desiderata is used if $R\beta$ is expected to be near C. The closeness of this approximate relation is regulated by the weights Γ. The inequality restrictions belong to the field of quadratic programming and are not discussed here (see e.g. Väliaho [6]).

3.2. Desiderata

When estimating parameters by the weighted least squares method the desiderata are treated in the same way as observations. We now have the model A^*

$$Y^* = X^*\beta^* + \varepsilon^*$$

based on the material

$$M^* = [W^*, X^*, Y^*] = \left[\overline{|W, \Gamma|}, \begin{bmatrix} X \\ R \end{bmatrix}, \begin{bmatrix} Y \\ C \end{bmatrix}\right]$$

with ε^* standing for $\begin{bmatrix} \varepsilon \\ \Phi \end{bmatrix}$. The principle of estimation is the same as in (2.2.2). Formulae (2.2.3) are now valid with B, E, Q, W, X, Y starred. On the other hand, formulae (2.4.2) and (2.5.1) have no meaning if there are desiderata in the model.

3.3. Equality restrictions

If the regression coefficients of a model $Y = XB + E$ are subject to the restrictions $RB = C$, these coefficients are determined so that, for $j = (k + 1), \dots, m$, (2.2.2) is minimum with respect to $\bar{b}_j$, subject to $R\bar{b}_j = c_j$. This problem may be solved by the LAGRANGE multiplier technique, cf. e.g. GOLDBERGER [2, pp. 45–8 and 256–8]. Denoting the unrestricted model by A and the restricted model by A^* we find the estimator

$$B^* = B + UR'(RUR')^{-1}(C - RB)\,. \tag{1}$$

Under the normality assumption, the hypothesis $R\beta_j = c_j$ can be tested against the alternative hypothesis $R\beta_j \neq c_j$; see GOLDBERGER [2, pp. 172–8].

However, we make another approach. If, in the weighting matrix $\Gamma = |\gamma_1 \cdots \gamma_p|$ of the corresponding desiderata, we let $\gamma_i \to \infty$ for some i these desiderata are converted into equality restrictions. The formulae for equality restrictions are thus obtained as limiting cases of the corresponding formulae for desiderata.

Finally, we note that, in the case of equality constraints,

$$s_j^2 = \frac{q_{jj}^*}{n + p - k}\,, \qquad j = (k + 1), \dots, m\,, \tag{2}$$

is an unbiased estimator of σ_j^2, and

$$\operatorname{cov}(b_j) = \sigma_j^2 U^*\,, \qquad j = (k + 1), \dots, m\,, \tag{3}$$

cf. formulae (2.4.2) and (2.5.1) in the case of unrestricted regression analysis.

4. Stepwise Regression Analysis

4.1. Fundamentals

Consider the elements a_{ij} in (2.3.3), with varying material and indicator matrix, as placed in the cells of a table (or in the storage of a computer). We shall refer to the variable content of this table as the model scheme, denoted by A.

In the stepwise method, certain transformations are applied in succession to the model scheme. One step constitutes the processing of one observation or variable. By processing an observation is meant a change in the weight of this observation. This also covers adding or removing an observation, since in the former case the observation can be considered to have been included in the original model with zero weight. Processing an observation will also be called a Gaussian step. By processing one (independent) variable we mean adding it to or removing it from regression.

In the stepwise method we shall call the variables in regression regressors and the variables not in regression regressands. More specifically, the classification is:

Regressors:

- the independent variables in regression,

Regressands:

- the dependent variables,
- the independent variables not in regression.

Thus the independent variables not in regression may technically be considered as being explained by means of the independent variables in regression at that stage. Processing a variable may now be expressed by saying that the independent variable concerned is shifted from the set of regressands to the set of regressors or vice versa. At a given stage of the model scheme, the rows and columns corresponding to the independent variables not in regression are usually not needed. They merely contain auxiliary quantities which make it possible to add these variables to regression later on. In each step, these quantities are updated.

Assume then that we have a fixed collection $\bar{Z}_1, \dots, \bar{Z}_h$ of observations. Let, in the Gaussian step i, $i = 1, \dots, n$, the weight of the observation $Z_i = \bar{Z}_{j_i}$, $1 \leqq j_i \leqq h$, be changed by the amount w_i (initially $\bar{Z}_1, \dots, \bar{Z}_h$ are considered to have zero weights). We define the data-pieces P_i, $i = 1, \dots, n$, by $P_i = [w_i, Z_i]$. Processing an observation can now be expressed by saying that the corresponding data-piece is added to the model. The material $M = [W, Z]$ is now defined as consisting of the data-pieces $P_1, \dots, P_n$. In addition, the notation $M_I = [W_I, Z_I]$, $I \subset \{1, \dots, n\}$, is introduced for a submaterial of M obtained from $[W, Z]$ by deleting the rows and columns $\bar{I}$ of W and the rows $\bar{I}$ of Z ($\bar{I}$ is the complement of I with respect to $\{1, \dots, n\}$).

Single stepping has been described above. In the sequel we shall first derive formulae for transforming the matrix A in the more general case of multiple stepping; formulae for single stepping are then obtained as particular cases.

4.2. Processing variables and observations

First we note that, starting from $Z'WZ$, variables may be processed by pivotal operations. Thus a variable z_i can be shifted from the set of regressands to the set of regressors, or vice versa, by $\mathscr{P}_i$. It is to be understood, that changing the sign of ψ_i is included in $\mathscr{P}_i$, cf. (2.3.2).

Then we turn to the processing of observations. Let A be a model based on the material $M_I = [W_I, X_I, Y_I]$, where I indicates the set of observations included in the material. Our purpose is to change the weights of the observations $[X_J, Y_J]$ by the amounts W_J or, in other words, to add the material $M_J = [W_J, X_J, Y_J]$ to A. We construct the matrix

$$\bar{A} = \begin{bmatrix} X_I'W_IX_I & X_I'W_IY_I & X_J' \\ Y_I'W_IX_I & Y_I'W_IY_I & Y_J' \\ -X_J & -Y_J & W_J^{-1} \end{bmatrix} \tag{1}$$

The desired model, say A^+, is obtained from $\bar{A}$ by $\mathscr{P}_{[1]}\mathscr{P}_{[3]}$ and deleting the last super-row and column. But now $\mathscr{P}_{[1]}\mathscr{P}_{[3]}\bar{A} = \mathscr{P}_{[3]}(\mathscr{P}_{[1]}\bar{A})$, where $\mathscr{P}_{[1]}\bar{A}$ is, except for augmentation, equal to A. Thus

$$\begin{aligned} U^+ &= U - UX_J'G_J^{-1}X_JU\,, \\ B^+ &= B + UX_J'G_J^{-1}(Y_J - X_JB)\,, \\ Q^+ &= Q + (Y_J - X_JB)'\,G_J^{-1}(Y_J - X_JB)\,, \end{aligned} \tag{2}$$

where $G_J = W_J^{-1} + X_J U X'_J$. Consequently, the addition of a data-piece $P_\lambda = [w_\lambda, Z_\lambda]$ to model A is performed by

$$a_{ij}^{+} = a_{ij} + \psi_j \frac{d_i d_j}{g} \quad \text{for all } i, j , \tag{3}$$

where

$$d_i = (1 - \varkappa_i)\, z_{\lambda i} + \sum_{j=1}^{n} \varkappa_j a_{ij} z_{\lambda j} ,$$

$$g = w_\lambda^{-1} + \sum_{j=1}^{n} \varkappa_j d_j z_{\lambda j}$$

with

$$\varkappa_i = \begin{cases} 0 & \text{if } \psi_i = 1 , \\ 1 & \text{if } \psi_i = -1 . \end{cases}$$

Finally, we note that, in a model scheme, variables and observations may be processed sequentially in any order.

4.3. The extended scheme

We proceed to construct an extended scheme, with one row and column associated with each variable and data-piece. It contains the model scheme as a part. It will appear that in this scheme, observations and variables can be processed simply by means of pivotal operations.

Suppose we are given a material $M = [W, Z]$ of order $n \times (n, m)$. It will prove convenient to number the data-pieces $(m + 1), \ldots$ $\ldots, (m + n)$. The extended scheme A is an $(m + n) \times (m + n)$ matrix, with the rows and columns corresponding to $z_1, \ldots, z_m$, $P_{m+1}, \ldots, P_{m+n}$, in this order. With any stage of the extended scheme we associate a corresponding indicator matrix

$$\Psi = \left|\overline{\psi_1} \cdots \underline{\psi_{m+n}}\right|$$

the correspondence between the ψ_j and the roles of the z_j and P_j being as follows:

Item	$\psi_j = -1$	$\psi_j = 1$
Variable $z_j, j = 1, \ldots, m$	regressor	regressand
Data-piece $P_j, j = (m + 1), \ldots, (m + n)$	not in the model	in the model

The stage of the extended scheme associated with a given indicator matrix Ψ is called an extended model.

We are now ready to construct the extended scheme explicitly. Having a material

$$M_{I+J} = \left[\left|\overline{W}_I, \underline{W_J}\right|, \begin{bmatrix} X_I \\ X_J \end{bmatrix}, \begin{bmatrix} Y_I \\ Y_J \end{bmatrix}\right]$$

we construct the matrix

$$A^{(0)} = \begin{bmatrix} 0 & 0 & X_I' & X_J' \\ 0 & 0 & Y_I' & Y_J' \\ -X_I & -Y_I & W_I^{-1} & 0 \\ -X_J & -Y_J & 0 & W_J^{-1} \end{bmatrix}$$

and put $\psi_1 = \cdots = \psi_m = 1$, $\psi_{m+1} = \cdots = \psi_{m+n} = -1$. We find, on denoting $A^{(1)} = \mathcal{P}_{[3]}A^{(0)}$ and $A^{(2)} = \mathcal{P}_{[1]}A^{(1)}$,

$$A^{(1)} = \begin{bmatrix} X_I'W_IX_I & X_I'W_IY_I & -X_I'W_I & X_J' \\ Y_I'W_IX_I & Y_I'W_IY_I & -Y_I'W_I & Y_J' \\ -W_IX_I & -W_IY_I & W_I & 0 \\ -X_J & -Y_J & 0 & W_J^{-1} \end{bmatrix}$$

and, denoting the model based on $[W_I, X_I, Y_I]$ by A,

$$A^{(2)} = \begin{bmatrix} U & B & -UX_I'W_I & UX_J' \\ -B' & Q & -E_I'W_I & E_J' \\ W_IX_IU & -W_IE_I & W_I - W_IX_IUX_I'W_I & W_IX_IUX_J' \\ X_JU & -E_J & -X_JUX_I'W_I & W_J^{-1} + X_JUX_J' \end{bmatrix} \tag{1}$$

with $E = Y - XB$. $A^{(2)}$ includes the variables X and the data-pieces P_i, $i \in I$.

We deduce that in the extended scheme, starting from $A^{(0)}$, variable z_j may be processed by $\mathcal{P}_j$, and data-piece P_j by $\mathcal{P}_j$. Again, changing the sign of ψ_j is included in $\mathcal{P}_j$.

In practice, we usually do not need all the rows and columns attached to the data-pieces. Therefore we often delete the unnecessary rows and columns of A, speaking still about the extended scheme.

If, in (1), the last super-row and column are decomposed into two super-rows and columns, corresponding to J_1 and J_2, it is observed that the super-row and column corresponding to J_2 may be calculated by means of the rest of $A^{(2)}$ and the submaterial $M_{J_2} = [W_{J_2}, X_{J_2}, Y_{J_2}]$. We find the following result: adding the row and column corresponding to an additional data-piece $[w_\lambda, Z_\lambda]$ to the (extended) model A of order l is accomplished by the formulae

$$\begin{aligned} a_{i,l+1} &= (1-\varkappa_i)\, z_{\lambda i} + \sum_{j=1}^{m} \varkappa_j a_{ij} z_{\lambda j}, && i = 1, \dots, l, \\ a_{l+1,j} &= -\psi_j a_{j,l+1}, && j = 1, \dots, l, \\ a_{l+1,l+1} &= w_\lambda^{-1} + \sum_{j=1}^{m} \varkappa_j a_{j,l+1} z_{\lambda j}, \end{aligned} \tag{2}$$

where

$$\varkappa_j = \begin{cases} 0 & \text{if } 1 \leqq i \leqq m \text{ and } \psi_i = 1, \\ 1 & \text{otherwise.} \end{cases}$$

These formulae are to be compared with (4.2.3); actually, (4.2.3) is equivalent to the augmentation of the model scheme A by (2), the performance of $\mathscr{P}_{l+1}$, and the deletion of row and column $l+1$.

Finally, it should be noted that in the extended scheme (and thus in the model scheme also) there is a special kind of redundancy:

$$a_{ij} = \psi_i \psi_j a_{ji}. \tag{3}$$

Thus it is sufficient to reserve a triangular matrix for the extended scheme (or model scheme).

4.4. Processing desiderata and equality restrictions

On the basis of their definition, desiderata may be processed (added, removed or their weights changed) by the same formulae as observations, replacing $[w_\lambda, X_\lambda, Y_\lambda]$ by $[\gamma_\lambda, R_\lambda, C_\lambda]$. Especially, this regards formulae (4.2.3). The augmentation is accomplished analogously to (4.3.2).

As for equality restrictions, formulae for processing them are obtained from the corresponding formulae for desiderata as limiting cases when letting the appropriate γ_i tend to infinity. Thus, for example, $G = RUR'$ in (4.2.2). Restrictions may be added analogously to (4.2.3), substituting $\gamma_\lambda^{-1} = 0$.

When applying the model scheme, equality restrictions cannot be removed because, from (4.2.2), $G^* = RU^*R' = 0$. Therefore, if equality restrictions are to be removed, the extended scheme must be used: when adding a restriction, the scheme is augmented by means of (4.3.2), and then the appropriate pivotal operation is performed.

To be specific, let A be a given stage of the model scheme. The addition of the restrictions $RB = C$, numbered $(m+1), \dots, (m+p)$, is accomplished by first augmenting A:

$$A = \begin{bmatrix} U & B & UR' \\ -B' & Q & (C - RB)' \\ RU & RB - C & G \end{bmatrix}$$

with $G = RUR'$ (letting $\psi_{m+1} = \cdots = \psi_{m+p} = -1$), and then applying $\mathscr{P}_{[3]}$:

$$A^* = \begin{bmatrix} U^* & B^* & -UR'G^{-1} \\ -(B^*)' & Q^* & (RB - C)'\,G^{-1} \\ G^{-1}RU & G^{-1}(RB - C) & G^{-1} \end{bmatrix}$$

Starting from A or A^*, any restriction i may be processed (added or removed) by $\mathscr{P}_i$. Moreover, any variable z_i may be processed by $\mathscr{P}_i$. and any data-piece $[w_\lambda, Z_\lambda]$ by formula (4.2.3) for $i, j = 1, \ldots, (m + p)$, defining $z_{\lambda i} = 0$ for $i > m$. If there is no need to remove a restriction $h > m + p$ afterwards, we may also use (4.2.3) for $i, j, = 1, \ldots$ $\ldots, (m + p)$. Variables, data-pieces, desiderata and equality restrictions may be processed in any order.

4.5. Discounting of observations [2]

In time series analysis it may be profitable to assign greater weights to recent observations than to older ones. This may be done by discounting: when adding observations one at a time, before each processing the weights of all the previous observations are multiplied by a discounting factor ϱ $(0 < \varrho < 1)$. Thus, if n observations are added subsequently, the observations are included in the final model with weights $\varrho^{n-1}w_1, \varrho^{n-2}w_2, \ldots, \varrho w_{n-1}, w_n$. The discounting factor may, of course, be changed during the process. It should be noted that discounting usually makes the variances of different observations unequal, thus causing the complications mentioned at the end of Sec. 2.5.

How does the discounting affect the extended scheme? We consider the matrix $A^{(2)}$ of (4.3.1). Before adding a new observation we must replace W_I by ϱW_I. This is accomplished by formulae

$$\begin{aligned} a_{ij}^+ &= \varrho^{-1} a_{ij} && \text{if } \psi_i = \psi_j = -1\,, \\ a_{ij}^+ &= a_{ij} && \text{if } \psi_i \neq \psi_j\,, \\ a_{ij}^+ &= \varrho a_{ij} && \text{if } \psi_i = \psi_j = 1\,, \end{aligned}$$

where the old and new matrices are denoted by A and A^+.

[2] See e.g. Törnqvist [3].

5. A Program for Stepwise Regression Analysis

5.1. Description of the program

GENERAL

The program has been written for the UNIVAC 1108 computer using FORTRAN V language. The name of the program is MTL * F.RA. The program is time sharing oriented. A batch processing oriented version, named MTL*F.RB, is also available (cf. Appendix 2, pp. 42-3).

INTRODUCTION

The regression analysis is based on observations on the variables $x_0 \equiv 1, x_1, \ldots, x_m$. Let $x_i, i \in I$, be independent variables and some others, $x_j, j \in J$, dependent variables. (x_0 is always considered as independent.) In addition, there may be a nonnegative weighting variable x_k, to be denoted w. The program computes weighted regression models of the form

$$x_j = \sum_{i \in I} b_{ij} x_i \qquad j \in J$$

assigning each observation h the weight w_h, or 1 if there is no weighting variable. It is to be understood that $b_{ij} = 0$ for every x_i, $i \in I$, not in regression. In addition, there may be constraints of the form

$$\sum_{i \in I} r_i b_{ij} = c_j , \qquad j \in J .$$

The program processes observations, independent variables and constraints, i.e. adds them to or removes them from a regression model, one at a time in any order. Several observations and variables may be processed by means of a single instruction. It is also possible to change the weight of an observation. Semi-automatic variable selection and discounting of observations are included in the program. If variable transformations are needed, the auxiliary transformation program MTL*F.TR is used.

METHOD

The method applied is for the most part after Väliaho [4]. A program run consists of operations on a matrix space in the computer, containing one row and column for each variable and constraint (actually, a triangular matrix space is used). Some methodological aspects are introduced below.

Degrees of freedom. Let n', m' and r' denote the numbers of observations, variables (including the possible constant term) and constraints included in a regression model. The number f of degrees of freedom is then given by

$$f = n' - m' + r'$$

Weighting. Weighting can be applied in two different cases:

(i) Classified data, with the weights giving the numbers of equal observations.

(ii) Non-classified data, when the weights are usually used for stabilizing the variances of different observations. If a noticeably different weighting is applied, then the estimators of the variances of the regression coefficients and the estimator of the residual variance are biased, and the t and F-tests are not valid, see VÄLIAHO [5].

Semi-automatic variable selection. At any stage the program prints, if desired, F-values, which indicate the significance of different independent variables for explanation. These F-values are approximate test values according to the F-distribution. Thus if the F-value attached to a variable in regression is small, it may be appropriate to remove the variable, and if the F-value attached to an independent variable not in regression is large, it may be profitable to add this variable. (Typical F-values are ≈ 4.) Thus the F-values provide a possibility of guiding the construction of regression models. This semi-automatic selection procedure is more flexible than an automatic procedure, cf. DRAPER and SMITH [1, pp. 167—73].

Discounting of observations. In time series analysis it may be profitable to assign greater weights to recent observations. This may be done by discounting: when adding observations one at a time, before each processing the weights of all the previous observations are multiplied by a disconuting factor ϱ $(0 < \varrho < 1)$. Thus, if n observations are added subsequently, the observations are included in the final model with weights $\varrho^{n-1}w_1, \varrho^{n-2}w_2, \ldots, \varrho w_{n-1}, w_n$. It is, however, possible to change the discounting factor during a program run. It should be noted that discounting usually makes the variances of different observations unequal, cf. case (ii) in the description of weighting above.

INPUT

As input material, there nay be

(i) an observation matrix
(ii) a moment matrix
(iii) a central moment matrix
(iv) file (i) and either (ii) or (iii)

The program requires that the appropriate file(s) be stored in the Fastrand storage.

Observation file. The observation file is read from punch cards or prepared by means of the transformation program MTL*F.TR. If the observation file is created by means of the transformation program, one need not look after the form of the file. If the file is created from punch cards, the DATA processor is made use of. The form of the file is as follows:

⟨header⟩
⟨format cards⟩
⟨observation matrix⟩

⟨header⟩ is in the form of a FORTRAN name list consisting of one or more cards as follows (a blanco column is indicated by ␣):

␣&HDR␣NAMES=⟨list⟩,
M = m ,
N = n ,
NF = s␣&END

where m, n and s are the numbers of variables, observations and format cards, respectively. The marked blank columns are obligatory; additional blank columns are allowed without any restrictions other than that &HDR and &END must be punched tightly, without blanks between, and that column 1 of continuation cards must be blank.

⟨list⟩ consists of m positive integers ($\leq$9999), separated by commas; they are the indices of the variables, the i:th index being attached to the i:th variable in the observation matrix. In ⟨list⟩, the following abbreviation may be used:

$$i, -j = i, i+1, \dots, j \qquad (i < j) .$$

In a ⟨list⟩ there may appear several abbreviations of this kind. The assumed form of NAMES=⟨list⟩ is NAMES=1, $-m$, it is to say tha if the specification of the indices is omitted, then the variables are num t bered 1, ... , m. The assumption of NF = s is NF = 1.

⟨format cards⟩ constitutes a FORTRAN format specification, which indicates the columns attached to the variables included in one observation, see Appendix 1, pp. 41—2.

⟨observation matrix⟩ is punched on the cards by rows so that one card contains values of variables attached to only one observation. On the other hand, an observation may be continued on several cards. The constant variable $x_0 \equiv 1$ need not be punched, because it is built into the program.

Moment files. The moment files are prepared by means of the transformation program, and one need not look after the form of the file.

INSTRUCTIONS

The instructions are divided into processing, printing and control instructions. An instruction may be continued on another row, by typing & at the end of the row. Below, the syntax structure or "grammar" of the instructions is introduced.

Processing instructions. The processing instructions, which affect the matrix space, are as follows:

AV␣⟨list⟩	add variable(s)
RV␣⟨list⟩	remove variable(s)
AO␣⟨list⟩	add observation(s)
RO␣⟨list⟩	remove observation(s)
CW␣h␣Δw␣Δf	change weight
AC␣h␣⟨coeff.⟩	add constraint
RC␣h	remove constraint

⟨list⟩ is defined on p. 30. In addition, the following additional abbreviation may now be used:

$$i, -j = i, i-1, \dots, j, \qquad (i > j).$$

The elements of ⟨list⟩ define the indices of the variables or observations to be processed.

In the instructions AV and RV, the index zero may be included in ⟨list⟩.

In the instruction CW, h stands for the index of an observation. Δw and Δf are the changes of the weight and the degrees of freedom respectively. It is to be noted that the operator must take care of Δf, because the program does not know whether, for example, a new observation is added or an old observation removed. However, Δf may be omitted; then the program assumes that $\Delta f = \Delta w$ in the case of classified material, and $\Delta f = 0$ otherwise.

In the instructions AC and RC, h is the index of a constraint (the indices are given by the operator). ⟨coeff.⟩ consists of pairs of numbers i, r_i or j, c_j; the pairs are separated by blanks. Here i and j are indices of variables, and r_i and c_j are decimal numbers (see p. 28). Only r_i, $c_j \neq 0$ are indicated. In the instruction AC, ⟨coeff.⟩ is omitted if the constraint concerned has been processed before.

Output instructions. There are two output instructions:

TP␣⟨specification⟩	temporary printing
PM␣⟨specification⟩	printing mode

TP affects output immediately according to the specification. PM stores the specification and affects the output after subsequent processing instructions (or their steps) until a new PM is given.

In an output specification, the following specifiers, separated by commas, may appear:

N	no output
R	multiple correlation coefficient and degrees of freedom
C	the preceding and the regression coefficients
CT	the preceding and the standard deviations of the regression coefficients, and the t-values
F	the F-values attached to the independent variables (the variables in regression are starred)
VA	analysis of variance
RES	residuals or predictive residuals (cf. the instruction PR below)
CORR	correlation matrix, or matrix of partial correlation coefficients adjusted for the independent variables in regression
MS	means and standard deviations of all variables, with the observations included in the model taken into account
MOM	the whole matrix space (a triangular matrix)

The output according to the seven first specifiers (N-RES) is given for every dependent variable.

Control instructions. The control instructions are as follows:

DV␣⟨list⟩	definition of dependent variables during a program run (⟨list⟩ is defined on p. 30)
DF␣ϱ	setting the discounting factor equal to ϱ
PR␣q	setting the number of predictive residuals
INIT	return the initial stage

LMM	return the initial stage and read the moment matrix
LOM	read the observation matrix
STOP	stopping the program run

In DF, the initial value is $\varrho = 1$.

If in PR, $q = 0$, the qualification RES affects the printing of all residuals. If $q > 0$, the specifier RES affects the printing of q nearest residuals after the observation just processed. If, however, the instruction was RO, the printing is started from the observation just processed. Initially, $q = 0$.

During the performance of INIT the observation file, if any, is preserved in the work area.

The instruction LMM destroys the observation matrix, if any, in the work area.

OPERATING

Starting the program. In case the program is run immediately after the transformation program, only the executive instruction

#XQT MTL*F.RA

is needed. Otherwise, before giving this instruction, the input files must be defined:

#ASG␣<observation file name>
#ASG␣<moment file name>
#USE␣10,<observation file name>
#USE␣11,<moment file name>

Of course, only the instructions attached to the available files are given. It appears, that the observation matrix is located on file no 10 and the moment matrix on file no 11. The observation file may also be typed or generated from paper tape (see ILLUSTRATION).

Definitions. At the beginning of a program run, the program makes some questions, which are to be replied:

Question 1: VARIABLES
Reply: <list>
Description: The variables introduced by <list> are taken to the analysis.

Question 2: WEIGHTING VARIABLE
Reply: (i) $\langle k \rangle$ or (ii) NONE
Description: (i) x_k is introduced as the weighting variable; (ii) there is no weighting variable.

Question 3: CLASSIFIED DATA
Reply: (i) YES or (ii) NO
Description: Question 3 is given only if there is a weighting variable. (i) classified data; (ii) non-classified data.

Question 4: MOMENT MATRIX
Reply: YES or NO
Description: The question is whether or not a moment matrix is to be used

Question 5: OBSERVATION MATRIX
Reply: YES or NO
Description: Cf. Question 4.

Question 6: DEPENDENT VARIABLES
Reply: (i) ⟨list⟩ or (ii) NONE
Description: (i) Definition of dependent variables; (ii) no dependent variables are defined at this stage.

Running the program. After the above definitions, the control moves to the operator. At any stage, the program asks

**INSTRUCTION XX

where XX is an integer. The desired instruction is typed. After performing any processing instruction the program gives output according to the current printing mode. In the case of the instructions AV, RV, AO and RO the printing is given after any step, i.e., after processing any variable/observation.

If there is a moment file (i.e. Reply 4 is YES) the appropriate part of the moment matrix constitutes the initial stage of the matrix space in the computer. Otherwise, the initial stage is a zero matrix. In this latter case the first processing instruction must be AO, resulting in the accumulation of a moment matrix. Initially, there are no independent variables in regression. However, if a program run is started from a central moment matrix, the constant term is included in the model at the beginning.

It should be noted that the set of dependent variables may be redefined during a program run by means of the instruction DV. The discounting factor may be changed by DF, and the number of predictive residuals by PR. The instructions INIT, LMM and LOM may also be given during a program run.

LIMITATIONS

The program requires that

$m + r \leqq 50\,, \qquad mn + m' \leqq 5000\,,$

where m is the number of the variables included in the analysis, m' is the number of the variables included in the observation matrix, r is the number of constraints, and n is the number of observations. It should be noted that the possible weighting variable is included in the observation matrix.

ERROR MESSAGES

In addition to the system error messages, the following error messages are possible:

No	Message	Note
1.	TOO MANY VARIABLES OR SYNTAX ERROR, RETYPE	Attached to Question 1 in OPERATING
2.	INCORRECT ANSWER, RETYPE	Attached to Question 2 or 6
3.	INCORRECT ANSWER, RETYPE YES OR NO	Attached to Question 3, 4 or 5
4.	NO CORRECT INPUT FILE	Both observation file and moment file are incorrect. The program is terminated
5.	OBSERVATION MATRIX CANNOT BE READ	One of messages 7—11, 13 follows
6.	MOMENT MATRIX CANNOT BE READ	One of messages 7, 8, 11, 12 follows
7.	ILLEGAL MATRIX	The file does not contain a matrix of correct form
8.	ILLEGAL HEADER	
9.	MORE THAN 6 FORMAT CARDS	
10.	TOO MUCH DATA	$mn + m' > 5000$, see LIMITATIONS
11.	UNAVAILABLE VARIABLES DESIRED	All desired variables are not found on a file
12.	TRANSMISSION ERROR	If a new trial is not successful, then the file in question has to be regenerated
13.	ERROR IN OBSERVATION XX	
14.	ACTUAL NUMBER OF OBSERVATIONS = XX	An observation matrix includes fewer observations than indicated in the header
15.	OBSERVATION MATRIX DESTROYED	Attached to the instruction LMM
16.	ILLEGAL OPERATION CODE	
17.	SYNTAX ERROR	
18.	VARIABLE XX CANNOT BE PROCESSED	One of the messages 21, 24, 25, 27, 28, 30 follows. The variables succeeding XX in ⟨list⟩ are not processed

No	Message	Note
19.	OBSERVATION XX CANNOT BE PROCESSED	One of messages 22, 23, 27, 30 follows. The observations succeeding XX in ⟨list⟩ are not processed
20.	CONSTRAINT XX CANNOT BE PROCESSED	One of messages 21, 24, 26—30 follows
21.	IMPROPER OPERATION CODE	Trial to add a variable in regression (or an imposed constraint), or to remove a variable not in regression (or an unimposed constraint)
22.	NONEXISTING OBSERVATION	Improper index of observation
23.	ILLEGAL WEIGHT	In CW, $\Delta w = 0$; or in AO or RO, $w_h < 0$
24.	NONEXISTING VARIABLE	
25.	TRIAL TO PROCESS A DEPENDENT VARIABLE	
26.	NONEXISTING CONSTRAINT	The constraint in question has not been processed before
27.	TOO FEW DEGREES OF FREEDOM	The number of degrees of freedom should become negative
28.	TOO MANY CONSTRAINTS REGARDING TO VARIABLES	The number of constraints should become greater than that of variables
29.	MATRIX SPACE LIMIT	The matrix space does not allow for a new constraint
30.	NONPOSITIVE PIVOT ELEMENT	The pivotal operation cannot be performed
31.	NO DEPENDENT VARIABLE DEFINED	
32.	ILLEGAL OUTPUT QUALIFICATION	
33.	CONSTANT TERM NOT INCLUDED	Attached to printing qualification VA
34.	NEGATIVE VALUE	In DF, $\varrho < 0$; or in PR, $q < 0$
35.	IMPROPER VARIABLE	Independent variable in DV

After any error message 4—35, a new instruction is questioned unless otherwise stated in the corresponding note. The instruction number is not increased except after messages 14, 15, and 33, which are of informative nature.

ILLUSTRATION

In the numerical example below, the data are taken from Draper and Smith [1, p. 116]. In the ordinary example, the input file is generated in time sharing from paper tape. It is shown separately how the input file is generated in a batch run.

```
#RUN RUN1,ACCOUNT,PROJ1
DATE: 092572        TIME: 110857
#PSW

PSW=8888888888
READY
#ASG,T FDATA,F2
READY
#DATA,I FDATA.
DATA005-09/25-11:10

**TAPE START**
  &HDR M=3,N=10,NF=1 &END
(F3.3,F3.1,F5.3)
353 20 1098
297 20 1113
308 23 1251
588 20  840
614 21  927
713 22  873
744 11  636
767 23  850
707 21  782
575 20  914

END OF TAPE
#END
#USE 10,FDATA.
READY
#XQT MTL*F.RA

  VARIABLES
1,-3

  WEIGHTING VARIABLE
NONE

  MOMENT MATRIX
NO
```

```
  OBSERVATION MATRIX
YES

  DEPENDENT VARIABLES
3

  ** INSTRUCTION NO        1
AO 1,-8

  ** INSTRUCTION NO        2
TP F

  VARIABLE    F-VALUE
     0          189.25
     1          22.886
     2          266.80

  ** INSTRUCTION NO        3
PM C,F

  ** INSTRUCTION NO        4
AV 2

  R          =     .53219
  D.F.       =          7

  VARIABLE     COEFF
     2         .46916

  VARIABLE    F-VALUE
     0         .95879
     1         1.6329
     2*        266.80

  ** INSTRUCTION NO        5
AV 1

  R          =     .58537
  D.F.       =          6

  VARIABLE     COEFF
     1        -.30842
     2         .54968
```

```
VARIABLE    F-VALUE
   0         64.381
   1*        1.6329
   2*        63.929

** INSTRUCTION NO      6
AV 0

R        =     .97116
D.F.     =          5

VARIABLE     COEFF
   0         .95791
   1        -.77113
   2         .20659

VARIABLE    F-VALUE
   0*        64.381
   1*        71.102
   2*        19.078

** INSTRUCTION NO      7
PM CT

** INSTRUCTION NO      8
AO 9,-10

ADDITION OF OBS        9

R        =     .96704
D.F.     =          6

VARIABLE     COEFF              STD DEV       T-VALUE
   0         .98979              .12174        8.1304
   1        -.80564              .90974-01    -8.8558
   2         .19687              .48706-01     4.0421

ADDITION OF OBS       10
R        =     .96818
D.F.     =          7
```

VARIABLE	COEFF	STD DEV	T-VALUE
0	.98917	.11275	8.7729
1	–.80580	.84323-01	–9.5562
2	.19691	.45149-01	4.3614

```
** INSTRUCTION NO      9
AC 1 0,2 1,.928 2,4 3,1.85

R      =     .86430
D.F.   =          8
```

VARIABLE	COEFF	STD DEV	T-VALUE
0	.73563	.20274	3.6285
1	–.58356	.14437	–4.0420
2	.23007	.89775-01	2.5627

```
** INSTRUCTION NO      10
STOP

END OF REGRESSION ANALYSIS
#FIN

RUNID: RUN1   ACCOUNT: ACCOUNT1   PROJECT: PROJ1
TIME: 00:00:00.867   IN: 101   OUT: 0   PAGES 2
INITIATION TIME:   11:08:57-SEP 25,1972
TERMINATION TIME:   11:16:48-SEP 25,1972

                        ****LINE INACTIVE****
```

The control cards for the generation of the observation file a batch run are as follows:

No	Control card
1	#RUN RUN1,ACCOUNT1,PROJ1
2	#ASG,C FDATA,F2
3	#DATA,I FDATA.
	(the cards forming the observation file)
4	#END
5	#FREE FDATA.
6	#SAVE PROJ1*FDATA.(30)
7	#FIN

If in card 3, I is replaced by IL, the cards forming the observation file are listed in output. The number 30 in card 6 indicates, that the file will be saved for 30 days.

APPENDIX 1: DESCRIPTION OF ⟨format cards⟩

In an observation file, ⟨format cards⟩ constitutes a FORTRAN format specification, which indicates the location on the cards of the variables included in one observation.

The columns of the cards attached to one observation are grouped into fields so that each variable has its own location (field) on the cards. The fields are defined from left to right using the following field specifiers:

Fw.d	a decimal number
Ew.d	a floating point number
wX	the field does not contain any data to be read

where

w	is an integer, which indicates the width of the field
d	is an integer, which indicates the number of decimal places (d has no effect if a decimal point has been punched in the field)

A format specification consists of field specifiers, which are separated by commas. However, if two successive fields are located on two distinct cards, they are separated by a slash (/). The first and last character of a specification are parentheses.

If there are p successive equal fields, the following abbreviation may be used:

pFw.d   or   pEw.d

Accordingly, if there are q successive equal field groups (a field group consists of several successive fields), the following abbreviation may be applied:

q(⟨field group specification⟩)

Finally, if the field specification for all cards attached to an observation are equal, it is sufficient to define only the first card.

A format specification may be located on most six cards, and only the columns 1-72 are allowable on each card.

Example. Suppose that an observation matrix, which includes 7 variables, has been punched on two cards as follows:

Variable	Card	Columns	Decimal places
1	1	1– 4	2
2	1	16–20	0
3	1	21–25	1
4	1	26–30	0
5	1	31–35	1
6	2	1– 5	3
7	2	6–10	3

An appropriate format specification is

(F4.2,11X,2(F5.0,F5.1)/2F5.3)

APPENDIX 2: BATCH PROCESSING ORIENTED VERSION

GENERAL

The name of the batch oriented version is MTL*F.RB. It differs from MTL*F.RA mainly at the following points: (i) The initial definitions are given in a different way; (ii) After an erroneous instruction, the program prints the appropriate error message and is terminated.

INITIAL DEFINITIONS

The variables, the weighting and the input files are defined on one or more cards which constitute a FORTRAN name list as follows:

␣ &RA␣PICK=⟨list⟩ ,
DV=⟨list⟩ ,
OM=s_1 ,
MM=s_2 ,
W=k ,
CL=t␣ &END

The variables to be considered are defined by means of the specifier PICK (cf. Question 1 on p. 33). The dependent variables are defined by means of DV (cf. Question 6 on p. 34). OM and MM are short for observation matrix and moment matrix. s_1 and s_2 are equal to 1 or 0 according to whether the appropriate input files are used or not (cf. Questions 4 and 5 on p. 34). The assumed values of s_1 and s_2 are zero. W and CL are needed only in the case of weighting. k is the index of the weighting variable (cf. Question 2 on p. 33), and t is equal to 1 or 0 according to whether or not the data material is classified. The assumed value of t is one (cf. Question 3 on p. 34).

INSTRUCTIONS

The instructions are punched on cards, one instruction on each card. In the remainder part of any card, a comment may be given; it must be separated from the instruction by a semicolon.

ILLUSTRATION

The example on pp. 37—40 may be carried through by means of MTL*F.RB as follows:

```
#RUN RUN1,ACCOUNT1,PROJ1
#PSW 8888888888
#ASG,T FDATA,F2
#DATA,I FDATA.
  &HDR M=3,N=10,NF=1 &END
(F3.3,F3.1,F5.3)
353 20 1098
297 20 1113
 :   :   :
575 20  914
#END
#USE 10,FDATA.
#XQT MTL*F.RB
  &RA PICK=1,-3,DV=1,OM=1 &END
AO 1,-8  ; ADDITION OF OBSERVATIONS
PM C,F   ; PRINTING MODE: REGR. COEFF. AND F-VALUES
AV 0,-2  ; ADDITION OF VARIABLES
PM CT    ; PRINTING MODE: REGR.COEFF. AND T-VALUES
AO 9,-10; ADDITION OF REMAINING OBSERVATIONS
AC 1 0,2 1,.928 2,4 3,1.85; ADDITION OF A CONSTRAINT
STOP     ; END OF ANALYSIS
#FIN
```

5.2. Technical notes

1° The structure of the program. The program consists of a controlling main program and a number of subprograms. The subprograms may be classified as follows:

(i) Definition
(ii) Communication with the operator
(iii) Transformation of the regression model
(iv) Output
(v) General subprograms

Subprograms in (i) initialize the program run and, during the program run, perform the instructions DV, DF, PR, INIT, LOM, and LMM.

Subprograms in (ii) ask for instructions, interpret them and give messages concerning possible errors.

Subprograms in (iii) constitute the core of the program. They transform variables which belong to the COMMON area, defining the regression model.

Subprograms (iv) calculate diverse estimates and print them.

2° **COMMON variables.** The COMMON area contains, in addition to the data-structures defining the regression model, the observation matrix, the external names[3] of the variables under consideration and a number of control variables.

At any stage of the program run, the matrix A of the model includes a row and a column for each variable under consideration, and each constraint processed thus far.

In addition, there is a row and a column for the constant variable $x_0 \equiv 1$ (the first row and column) and an auxiliary row and column for the quantities d_i, g (the last row and column), see (4.2.3).

Technically, the matrix of the model is saved in vector A in the form of a lower triangular matrix. The address of the first element of any row is saved in vector R so that the index i' of an element a_{ij} $(i \geqq j)$ in vector A is obtained from

$$i' = R(i) + j \, .$$

Any element a_{ij} $(i < j)$ is obtained, when needed, from a_{ji} using (4.3.3). The indicator numbers ψ_i are saved in vector K so that $K(i) = \psi_i$.

For the definition of the regression model, the following additional variables are used:

DVLIST	A logical vector, which defines the dependent variables
SX, SSX, SW	Sums and sums of squares of variables, and the sum of weights, calculated from the observations included in the model
OBS	Number of observations included in the model
DF	Number of degrees of freedom

The observation matrix is saved by rows in vector OMATR. If weighting is used, the observation matrix is augmented by a column for weights. The external names of the variables under consideration are

[3] External names are to the ordinary names of variables, used by the operator. Internal names are technical names, used by the program.

saved in vector NAME so that the name of the variable attached to column i in the observation matrix is saved in NAME(i). An observation to be considered is shifted into the COMMON vector X. The control variables are as follows:

INDEP	Number of independent variables included in the model
DEPS	Number of dependent variables included in the model
CONSTS	Number of imposed constraints
CONS	Number of constraints included in the matrix of the model
NOBS	Number of observations included in the observation matrix
VARS	Number of variables included in the observation matrix (the possible weighting variable is not included)
NW	Number of weighting variables (NW = 0 or 1)
HOMOG	In the case of weighting, indicates whether or not the material is classified
DISC	Discounting factor
NPR	Number of predictive residuals desired
LOBS	Indicator of the base observation when calculating predictive residuals
ERROR	Error indicator

3° **List of programs.** The main program is as follows:

00 MAIN Control

The subprograms are as follows:

(i) Definition

01	DEF	Definition of COMMON variables
02	INIT	Asking for the initial definitions (pp. 33–4)
03	IVALS	Initialization of COMMON variables
04	GETOM	Reading the observation matrix (only the variables under consideration are taken into account)
05	GETMOM	Reading the moment matrix (the whole moment matrix is first read into the space for the observation matrix, and then the appropriate part is shifted into the vector A)
06	DEPLIS	Definition of the dependent variables
07	CHGIND	Transforming the external name of a variable into an internal name with the aid of vector NAME
08	SETNPR	Setting the number of predictive residuals desired
09	SETDC	Setting the discounting factor

(ii) Communication with the operator

10	GETINS	Asking for the following instruction, checking the syntax of it, and interpretation of the instruction
11	TABIN	Separating an instruction from a card
12	CHAR	Separating one character from a card
13	ERRMSG	Giving error messages

(iii) Transformation of the regression model

20	SERINS	Controlling the instructions AV, RV, AO, and RO, and giving output under the instruction PM (the ordinary transformation is performed under subprograms 21—24)
21	HANVAR	Processing a variable
22	HANOBS	Processing an observation
23	HANCON	Processing a constraint
24	CHGWEI	Changing a weight
25	PIVOT	Performance of a pivotal operation on the matrix A
26	GAUSS	Performance of a Gaussian step
27	DISCON	Discounting of a regression model
28	UPDATE	Updating SX, SSX and SW when processing an observation
29	SANDR	Shifting an observation into the vector X

(iv) Output

30	CHGOUT	Interpretation of the specification TP or PM (the result is an internal output code)
31	OUTPUT	Controlling the output
32	OUTP1	Giving the general output
33	OUTP2	Giving the output for a dependent variable
34	FVALUE	Calculating the F-values

(v) General subprograms

40	OPENL	Interpreting a list
41	CHGPL	Constructing a pick list, with the aid of which the variables to be considered are picked from the observation matrix or from the moment matrix

5.3. Program listing

```
C 00
C MAIN PROGRAM
      INCLUDE DEF,LIST
      INTEGER CODE,OCODE,PCODE,NP,VARS1
      REAL    PLIST(P3)
      LOGICAL    FATERR
      EXTERNAL HANVAR,HANOBS,DEPLIS
9000  FORMAT ('0','END OF REGRESSION ANALYSIS')
9001  FORMAT ('1STEPWISE REGRESSION',
     1'    MTL*F.RA'/2X,37('=')//)
      WRITE (6,9001)
      CALL INIT (FATERR)
      IF (FATERR) GO TO 13
110   CALL GETINS (CODE,PLIST,NP,ERROR)
      ERROR=.FALSE.
C EXECUTE THE INSTRUCTION
100   GO TO (1,1,3,3,5,5,7,8,9,10,11,12,13,14,15,16), CODE
13    WRITE(6,9000)
      STOP
1     CALL SERINS(CODE,PLIST,NP,HANVAR,PCODE)
      GO TO 110
3     CALL SERINS(CODE,PLIST,NP,HANOBS,PCODE)
      GO TO 110
5     CALL HANCON (CODE,PLIST,NP)
      GO TO 120
7     CALL CHGWEI (PLIST,NP)
      GO TO 120
8     VARS1=VARS+1
      DO 58 I=1,VARS1
58    DVLIST(I)=.FALSE.
      DEPS=0
      CALL SERINS(CODE, PLIST, NP,DEPLIS,0)
      GO TO 110
9     CALL CHGOUT (OCODE,PLIST,NP,ERROR)
      IF (.NOT.ERROR) PCODE=OCODE
      GO TO 110
10    CALL CHGOUT (OCODE,PLIST,NP,ERROR)
      IF (.NOT.ERROR) CALL OUTPUT (0,0,OCODE)
      GO TO 110
11    CALL IVALS
      PCODE=0
      GO TO 110
```

```
12     CALL GETOM(ERROR)
       GO TO 110
14     CALL SETDC (PLIST, NP)
       GO TO 110
15     CALL GETMOM (ERROR)
       GO TO 110
16     CALL SETNPR (PLIST,NP)
       GO TO 110
C OUTPUT AFTER HANDLING A CONSTRAINT OR
C CHANGING WEIGHT
120    IF (ERROR) GO TO 110
       IF (PCODE.NE.0) CALL OUTPUT (0,IFIX(PLIST(1)),PCODE)
       GO TO 110
       END
C 01
DEF PROC
C GLOBAL VARIABLES AND DEFINITIONS
       PARAMETER   P0=52,P1=(P0+1)*P0/2,P2=5000,IN=5,OUT=6
      1            ,P3=2*P0
       COMMON      A(P1),K(P0),X(P0),SX(P0),SSX(P0),SW,
      1            OMATR(P2),DF,OBS,NOBS,VARS,INDEP,
      2            DVLIST(P0),CONSTS,CONS,HOMOG,WIND,
      3            NAME(P0),R(P0),ERROR,NW,DISC,LOBS,NPR,
      4            DEPS
       INTEGER     K,DF,OBS,NOBS,VARS,INDEP,CONSTS,CONS,
      1            WIND,NAME,R,NW,LOBS,NPR,DEPS
       REAL        A,SX,SSX,SW,OMATR,DISC
       LOGICAL     DVLIST,HOMOG,ERROR
  END
C 02
       SUBROUTINE INIT (FATERR)
C INITIALIZE DATA STRUCTURES
       INCLUDE DEF,LIST
       INTEGER     NUMER/2/,TYPE(P0),O23/O23/,O36/O36/
       LOGICAL     ERR,ERR1,FATERR
       REAL        PLIST(P0)
       EXTERNAL    DEPLIS
9020   FORMAT ('0VARIABLES')
9021   FORMAT('0WEIGHTING VARIABLE')
9022   FORMAT ()
9023   FORMAT ('0CLASSIFIED DATA')
9024   FORMAT (A6)
9025   FORMAT (' TOO MANY VARIABLES OR SYNTAX ERROR',
      1' , RETYPE')
9026   FORMAT ('0MOMENT MATRIX')
```

```
9027  FORMAT ('0OBSERVATION MATRIX')
9028  FORMAT ('0DEPENDENT VARIABLES')
9030  FORMAT (' INCORRECT ANSWER, RETYPE YES OR NO')
9031  FORMAT (' NO CORRECT INPUT FILE')
9033  FORMAT (' INCORRECT ANSWER, RETYPE')
C INITIAL VALUES
      CALL IVALS
      DO 1600 I=1,P0
1600  R(I)=I*(I-1)/2
      NOBS=0
      VARS=0
      NW=0
C VARIABLE LIST
      WRITE (OUT ,9020)
1640  ERR=.FALSE.
      CALL TABIN (PLIST,TYPE,NP,51,ERR)
      IF (.NOT.ERR) GO TO 1645
         WRITE (OUT,9025)
         GO TO 1640
1645  IF (NP.EQ.0) GO TO 1690
C CHECK SYNTAX
      DO 1630 I=1,NP
          IF (TYPE(I).EQ.NUMER) GO TO 1630
              WRITE (OUT,9025)
              GO TO 1640
1630  CONTINUE
      NAME(1)=0
      CALL OPENL (PLIST,NP,NAME,2,VARS,ERR,P0)
      IF (.NOT.ERR) GO TO 1690
         WRITE (OUT,9025)
         GO TO 1640
C WEIGHTING
1690  WRITE (OUT, 9021)
1691  ERR=.FALSE.
      CALL TABIN (PLIST,TYPE,NP,2,ERR)
      IF (NP.EQ.1.AND..NOT.ERR) GO TO 1692
1693    WRITE (OUT,9033)
        GO TO 1691
1692  IF (FLD(0,6,PLIST(1)).EQ.O23) GO TO 1650
         IF (TYPE(1).NE.2) GO TO 1693
      WIND=PLIST(1)
      NW=0
      IF (WIND.LT.1) GO TO 1650
      NW=1
      WRITE (OUT,9023)
```

```
1670  ERR=.FALSE.
      CALL TABIN (PLIST,TYPE,NP,2,ERR)
      IF (NP.EQ.1.AND.(FLD(0,6,PLIST(1)).EQ.O36.OR.FLD(0,6,
     1PLIST(1)).EQ.O23).AND..NOT.ERR) GO TO 1660
            WRITE (OUT,9030)
            GO TO 1670
1660  HOMOG=.TRUE.
      IF (FLD(0,6,PLIST(1)).EQ.O23) HOMOG=.FALSE.
C LOAD MOMENT AND OBSERVATION MATRICES
1650  WRITE (OUT,9026)
1755  ERR=.FALSE.
      CALL TABIN (PLIST,TYPE,NP,2,ERR)
      IF (NP.EQ.1.AND.(FLD(0,6,PLIST(1)).EQ.O36.OR.FLD(0,6,
     1PLIST(1)).EQ.O23).AND..NOT.ERR) GO TO 1760
            WRITE (OUT,9030)
      GO TO 1755
1760  IF (FLD(0,6,PLIST(1)).EQ.O36) CALL GETMOM(ERR)
      WRITE (OUT,9027)
1765  ERR1=.FALSE.
      CALL TABIN (PLIST,TYPE,NP,2,ERR1)
      IF (NP.EQ.1.AND.(FLD(0,6,PLIST(1)).EQ.O36.OR.FLD(0,6,
     1PLIST(1)).EQ.O23).AND..NOT.ERR1) GO TO 1770
           WRITE (OUT,9030)
           GO TO 1765
1770  IF (FLD(0,6,PLIST(1)).EQ.O36) CALL GETOM (ERR1)
      IF (.NOT.ERR.OR..NOT.ERR1) GO TO 1780
          FATERR=.TRUE.
          WRITE (OUT,9031)
          RETURN
C DEPENDENT VARIABLES
1780  WRITE (OUT,9028)
1785  ERR=.FALSE.
      CALL TABIN (PLIST,TYPE,NP,50,ERR)
      IF (.NOT.ERR) GO TO 1786
         WRITE (OUT,9033)
         GO TO 1785
1786  IF (FLD(0,6,PLIST(1)).EQ.O23) RETURN
      IF (NP.EQ.0) RETURN
C CHECK SYNTAX
      DO 1790 I=1,NP
           IF (TYPE(I).EQ.NUMER) GO TO 1790
               WRITE (OUT,9025)
               GO TO 1785
```

```
1790  CONTINUE
      CALL SERINS (8,PLIST,NP,DEPLIS,0)
      RETURN
      END
C 03
      SUBROUTINE 1VALS
C INITIALIZE VALUES OF COMMON VARIABLES AND TABLES
      INCLUDE  DEF,LIST
      INTEGER I
      DO 3500 I=1,P0
         K(I)=1
         SX(I)=0.
         SSX(I)=0.
3500     DVLIST (I)=.FALSE.
      DO 3510 I=1,P1
3510     A(I)=0.
      SW=0.
      DF=0
      OBS=0
      INDEP=0
      DEPS=0
      CONSTS=0
      CONS=0
      PCODE=0
      LOBS=0
      NPR=0
      DISC=1.
      RETURN
      END
C 04
      SUBROUTINE GETOM (ERR)
C READ THE OBSERVATION MATRIX
      INCLUDE DEF,LIST
      PARAMETER P10=2*P0,F10=10,FL=84
      INTEGER  FID,M,N,NAMES1(P10),NF,PLIST(P0),UPPER,ERNR,
     1         ERIND,NVAR,I,J,J1,J2,J3,CL,NN,NN1
      LOGICAL  ERR
      REAL     FORM(FL)/84*'    '/,NAMES(P10)
      NAMELIST /HDR/FID,N,M,NAMES,NF,CL
9030  FORMAT (12A6)
      ERR=.FALSE.
      FID=1
      NF=1
C READ THE HEADER
      DO 1600 I=1,P10
```

```
1600  NAMES(I)=0
      READ (F10,HDR,ERR=1620,END=1620)
      IF (NAMES (1).NE.0) GO TO 1500
          DO 1520 I=1,M
               NAMES1(I)=I
1520      CONTINUE
          GO TO 1530
1500  NN=0
      DO 1510 I=1,P10
           IF (NAMES(I).EQ.0) GO TO 1540
               NN=NN+1
1510  CONTINUE
1540  CALL OPENL (NAMES,NN,NAMES1,1,NN1,ERR,P10)
      IF (ERR.OR.NN1.NE.M) GO TO 1620
1530  IF (FID.EQ.1) GO TO 1601
          ERNR=14
          GO TO 1690
1601  IF (VARS.NE.0) GO TO 1602
          IF (M.LE.P0-2) GO TO 1590
              ERNR=15
              GO TO 1690
1590      DO 1595 I=1,M
1595           NAME (I+1)=NAMES1 (I)
          NAME(1)=0
          VARS=M
1602  NVAR=VARS
      IF (NW.EQ.0) GO TO 1605
          NVAR=VARS+NW
          NAME(NVAR+1)=WIND
C CONSTRUCT THE PICK LIST
1605  CALL CHGPL (NAME,NAMES1,PLIST,NVAR,M,ERR)
      IF (.NOT.ERR) GO TO 1607
          ERNR=19
          GO TO 1690
C CHECK IF THERE IS TOO MUCH DATA
1607  IF (NVAR*N+M.LE.P2) GO TO 1608
          ERROR=.TRUE.
          ERNR=16
          GO TO 1690
C READ THE FORMAT DEFINITION
1608  IF (NF.LE.6) GO TO 1606
          ERNR=17
          GO TO 1690
1606  UPPER=NF*12
      READ (F10,9030,ERR=1610,END=1610)(FORM(I),I=1,UPPER)
```

```
C READ THE OBSERVATION MATRIX
      J1=1
      DO 1650 I=1,N
      J2=J1+NVAR
      J3=J2+M-1
      READ (F10,FORM,ERR=1660,END=1670)(OMATR(J),J=J2,J3)
      DO 1680 J=1,NVAR
      J3=J2+PLIST(J)-1
1680  OMATR(J1+J-1)=OMATR(J3)
1650  J1=J2
      NOBS=N
      REWIND F10
      RETURN
C ERROR EXIT
1610  ERNR=17
      ERIND=0
      GO TO 1690
1620  ERNR=18
      ERIND=0
      GO TO 1690
1660  ERNR=40
      ERIND=(J1-1)/NVAR
      GO TO 1690
C NUMBER OF OBSERVATIONS IS LESS THAN STATED
1670  ERNR=41
      NOBS=(J1-1)/NVAR
      CALL ERRMSG (ERNR,NOBS,0)
      GO TO 1710
1690  ERROR=.TRUE.
      ERR=.TRUE.
1700  CALL ERRMSG(ERNR,ERIND,4)
1710  REWIND F10
      RETURN
      END
C 05
      SUBROUTINE GETMOM (ERR)
C READ THE MOMENT MATRIX
      INCLUDE DEF,LIST
      PARAMETER P10=2*P0,F11=11
      INTEGER FID,NAMES(P10),PLIST(P0),ERNR,CL
      LOGICAL ERR,CENT
      NAMELIST /HDR/FID,N,M,NAMES,NF,CL
9005  FORMAT (10A6)
      ERR=.FALSE.
```

```
C READ THE HEADER
      READ (F11,HDR,ERR=2000,END=2000)
      IF (FID.EQ.2.OR.FID.EQ.3) GO TO 2020
         ERNR=14
         GO TO 2030
2020  IF (VARS.NE.0) GO TO 2040
         IF (M.LE.P0-2) GO TO 2050
         ERNR=15
         GO TO 2030
2050     DO 2060 I=1,M
2060     NAME (I+1)=NAMES (I)
         NAME(1)=0
         VARS=M
C INDICATE IF OBSERVATION MATRIX HAS BEEN READ BEFORE
2040  IF (NOBS.EQ.0) GO TO 2070
         CALL ERRMSG (22,0,0)
         NOBS=0
C COMPUTE THE PICKLIST
2070  CALL CHGPL (NAME, NAMES, PLIST, VARS, M, ERR)
      IF (.NOT.ERR) GO TO 2075
         ERNR=19
         GO TO 2030
C READ SW,SX, SSX
2075  READ (F11,9005,ERR=2080,END=2080) SW
      READ (F11, 9005, ERR=2080, END=2080) (OMATR(I), I=1,M)
      CALL PICK (SX)
      READ (F11,9005,ERR=2080,END=2080) (OMATR(I),I=1,M)
      CALL PICK (SSX)
C READ THE MOMENT MATRIX AND CONSTRUCT MATRIX A
      MSIZE=(M+1)*M/2
      IF (MSIZE.LE.P2) GO TO 2076
         ERNR=16
         GO TO 2030
2076  READ (F11,9005,ERR=2080,END=2080) (OMATR(I),I=1,MSIZE)
      CENT=.FALSE.
      IF (FID.EQ.3) CENT=.TRUE.
      A(1)=SW
      IF (CENT) A(1)=1./SW
      IJ=2
      DO 2100 I1=1,VARS
      I=PLIST(I1)
      A(IJ)=SX(I1+1)
      IF (CENT) A(IJ)=-A(IJ)/SW
      IJ=IJ+1
```

```
      DO 2090 J1=1,I1
      J=PLIST(J1)
      IF (J.LE.I) IM=I*(I-1)/2+J
      IF (J.GT.I) IM=J*(J-1)/2+I
      A(IJ)=OMATR(IM)
      IJ=IJ+1
2090  CONTINUE
2100  CONTINUE
      IF (CL.EQ.1) N=SW
      OBS=N
      DF=N
      IF (.NOT.CENT) GO TO 2110
         K(1)=-1
         INDEP=1
         DF=DF-1
2110  REWIND F11
      RETURN
C ERROR EXIT
2000  ERNR=18
      GO TO 2030
2080  ERNR=20
2030  CALL ERRMSG (ERNR,0,5)
      ERR=.TRUE.
2010  RETURN
      SUBROUTINE PICK (X)
C PICK SX OR SSX
      REAL       X (P0)
      DO 2200 I=1,VARS
         I1=PLIST (I)
2200     X(I+1)=OMATR(I1)
      RETURN
      END
C 06
      SUBROUTINE DEPLIS (CODE,IND)
C DEFINE THE EXPLICIT DEPENDENT VARIABLES
      INCLUDE DEF,LIST
      INTEGER        CODE,IND,ERNR,CHGIND,INT
C CHECK THE INSTRUCTION
      INT=CHGIND(IND)
      IF (INT.GT.0) GO TO 600
         ERNR=6
         GO TO 610
600   IF (K(INT).GT.0) GO TO 620
         ERNR=24
C ERROR EXIT
```

```
610   ERROR=.TRUE.
      CALL ERRMSG(ERNR,IND,0)
      RETURN
620   DVLIST(INT)=.TRUE.
      DEPS=DEPS+1
      RETURN
      END
C 07
      INTEGER FUNCTION CHGIND(IND)
C TRANSFORM THE EXTERNAL NAME TO INTERNAL
      INCLUDE     DEF,LIST
      INTEGER     VARS1,IND,I
      VARS1=VARS+1
      DO 1100  I=1,VARS1
           IF (NAME(I).NE.IND) GO TO 1100
                CHGIND=I
                GO TO 1110
1100  CONTINUE
C VARIABLE NOT DEFINED
      CHGINC=-IND
1110  RETURN
      END
C 08
      SUBROUTINE SETNPR (PLIST,NP)
C SET THE NUMBER OF PREDICTIVE RESIDUALS
      INCLUDE DEF,LIST
      INTEGER NP,ERNR
      REAL PLIST (NP)
      IF (NP.EQ.1) GO TO 1000
         ERNR=23
         GO TO 1010
1000  IF (PLIST(1).GE.0) GO TO 1020
         ERNR=21
C ERROR EXIT
1010  ERROR=.TRUE.
      CALL ERRMSG(ERNR,0,0)
      RETURN
1020  NPR=PLIST(1)
      RETURN
      END
C 09
      SUBROUTINE SETDC (PLIST,NP)
C SET THE DISCOUNTING FACTOR
      INCLUDE   DEF,LIST
      INTEGER NP,ERNR
```

```
      REAL      PLIST(NP)
      IF (NP.EQ.1) GO TO 1010
          ERNR=23
          GO TO 1020
1010  IF (PLIST (1).GT.0) GO TO 1000
          ERNR=21
C ERROR EXIT
1020  ERROR=.TRUE.
      CALL ERRMSG(ERNR,0,0)
      RETURN
1000  DISC= PLIST (1)
      RETURN
      END
C 10
      SUBROUTINE GETINS (CODE,PLIST,NP,ERROR)
C GET INSTRUCTION
      PARAMETER  MAXCOD=16,MAXL=104,ALFA=1,NUMER=2,
     1           OUT=6
      INTEGER    NP,INSNR/0/,CODE,TYPE,I,PTYPE
      REAL       CODE1
      LOGICAL    ERROR,ERR
      DIMENSION  PLIST(MAXL),TYPE(MAXL),CODTAB(MAXCOD)
      DATA       CODTAB /'AV','RV','AO','RO','AC','RC','CW'
     1           ,'DV','PM','TP','INIT','LOM','STOP','DF'
     2           ,'LMM','PR'/
9002  FORMAT (' '//' ** INSTRUCTION NO',I6)
9003  FORMAT (' ILLEGAL OPERATION CODE ')
9004  FORMAT (' SYNTAX ERROR ')
C GET INSTRUCTION LIST
      IF (.NOT.ERROR)INSNR=INSNR+1
800   WRITE(6,9002) INSNR
      ERR=.FALSE.
      CALL TABIN (PLIST,TYPE,NP,MAXL,ERR)
      IF (.NOT.ERR) GO TO 810
          WRITE (OUT,9004)
          GO TO 800
C INTERPRET CODE
810   CODE1=PLIST(1)
      DO 820 I=1,MAXCOD
      IF (CODE1.NE.CODTAB(I)) GO TO 820
          CODE=I
          GO TO 830
820   CONTINUE
```

```
C UNDEFINED CODE
      WRITE(6,9003)
      GO TO 800
C CHECK SYNTAX AND MODIFY PLIST
830   PTYPE=NUMER
      IF (CODE.EQ.10.OR.CODE.EQ.9) PTYPE=ALFA
      IF (NP.LT.2) GO TO 850
         DO 840  I=2,NP
               IF (TYPE(I).EQ.PTYPE) GO TO 835
                  WRITE (OUT,9004)
                  GO TO 800
835              PLIST(I-1)=PLIST(I)
840      CONTINUE
850   NP=NP-1
      RETURN
      END
C 11
      SUBROUTINE TABIN(TAB,TYPE,N,NU,ERROR)
C READ THE INSTRUCTION LIST
      PARAMETER COLS=72,IN=5,OUT=6
      DIMENSION TAB(NU),A(14),F1(3),F2(3)
      INTEGER    TYPE(NU),COL,FIRST,TYP
      LOGICAL   ERROR
100   FORMAT (13A6,A2)
101   FORMAT ('(T',I2,',A',I2,')')
102   FORMAT ('(T',I2,',G',I2,'.0)')
103   FORMAT (4X,12A6,A3)
      N=0
C READ A CARD
10    READ(IN,100,END=50,ERR=50) A
      COL=0
C SEEK THE FIRST CHARACTER OF LIST ELEMENT
12    COL=COL+1
      IF (COL.GT.COLS) RETURN
      CALL CHAR(A,COL,NCHAR)
15    IF (NCHAR.EQ.5.OR.NCHAR.EQ.46) GO TO 12
      IF (NCHAR.EQ.59) RETURN
      IF (CHAR.EQ.38) GO TO 10
      IF (N+1.GT.NU) GO TO 50
      FIRST=COL
      TYP=1
      IF (NCHAR.GE.33) TYP=2
      LENGTH=0
```

```
C SEEK THE END OF LIST ELEMENT
20    COL=COL+1
      LENGTH=LENGTH+1
      IF (COL.GT.COLS) GO TO 30
      CALL CHAR(A,COL,NCHAR)
      IF (NCHAR.EQ.5.OR.NCHAR.EQ.46.OR.NCHAR.EQ.38.OR.NCHAR.
     1EQ.59) GO TO 30
      IF(TYP.EQ.2.AND.(NCHAR.LT.48.OR.(NCHAR.GT.57.AND.NCHAR.
     1NE.61))) GO TO 50
      GO TO 20
C SEPARATE THE ELEMENT
30    IF(TYP.EQ.2) GO TO 35
      IF(LENGTH.GT.6) GO TO 50
      ENCODE(101,F1)FIRST,LENGTH
      DECODE (F1,A) TAB(N+1)
      GO TO 40
35    ENCODE(102,F2)FIRST,LENGTH
      DECODE (F2,A) TAB(N+1)
40    TYPE(N+1)=TYP
      N=N+1
C RETURN IF CARD ENDS
      IF (COL.LE.COLS) GO TO 15
      RETURN
50    ERROR=.TRUE.
      RETURN
      END
C 12
      SUBROUTINE CHAR(A,I,NCHAR)
C TAKE THE I TH CHARACTER FROM CARD
      PARAMETER CARD=14,P2=6*CARD
      DIMENSION A(CARD)
      INTEGER   I,NCHAR,J,K
      IF(I)10,10
      IF(I-P2)20,20
10    NCHAR=-1
      RETURN
20    J=(I-1)/6+1
      K=6*MOD(I-1,6)
      IF (K.EQ.0)K=0
      NCHAR=FLD(K,6,A(J))
      RETURN
      END
```

```
C 13
      COMPILER (DATA=IBM)
      SUBROUTINE  ERRMSG (NR,IND,HNR)
C ERROR AND DIAGNOSTIC MESSAGES
      INTEGER  NR, IND, HNR, SEEK(24),I1,I2,I
      REAL     HDR(10),MSG(99)
      DATA     HDR /'VARIAB','LE    ',
     1              'OBSERV','ATION ',
     2              'CONSTR','AINT  ',
     3              'OBSERV','ATION ',
     4              '    MO','MENT  '/
      DATA  MSG(1)  /'IMPROP','ER OPE','RATION',' CODE ',
     2              'TOO FE','W DEGR','EES OF',' FREED',
     2              'OM    ',
     3              'TOO MA','NY CON','STRAIN','TS REG',
     3              'ARDING',
     4              ' TO VA','RIABLE','S     ',
     5              'TRIAL ','TO PRO','CESS A',' DEPEN',
     5              'DENT V','ARIABL','E     ',
     6              'NONPOS','ITIVE ','PIVOT ','ELEMEN',
     6              'T     ',
     7              'NONEXI','STING ','VARIAB','LE    ',
     8              'NONEXI','STING ','OBSERV','ATION ',
     9              'ILLEGA','L WEIG','HT    '/
      DATA MSG(41)  /'NONEXI','STING ','CONSTR','AINT  ',
     2              'MATRIX',' SPACE',' LIMIT',
     3              'NO DEP','ENDENT',' VARIA','BLE DE',
     3              'FINED ',
     4              'ILLEGA','L OUTP','UT QUA','LIFICA',
     4              'TION  ',
     5              'CONSTA','NT TER','M NOT ','INCLUD',
     5              'ED    ',
     6              'ILLEGA','L MATR','IX    ',
     7              'TOO MA','NY VAR','IABLES',
     8              'TOO MU','CH DAT','A     ',
     9              'MORE T','HAN 6 ','FORMAT',' CARDS'/
      DATA MSG(76)  /'ILLEGA','L HEAD','ER    ',
     2              'UNAVAI','LABLE ','VARIAB','LES DE',
     2              'SIRED ',
     3              'TRANSM','ISSION','ERROR ',
     4              'NEGATI','VE VAL','UE    ',
     5              'OBSERV','ATION ','MATRIX',' DESTR',
     5              'OYED  ',
     6              'SYNTAX',' ERROR',
     7              'IMPROP','ER VAR','IABLE'/
```

```
      DATA      SEEK/O00104,O00505,O01210,O02207,O03105,O03604,
     1                 O04204,O04603,O05104,O05503,O06005,O06505,
     2                 O07205,O07703,O10203,O10503,O11004,O11403,
     3                 O11705,O12403,O12703,O13205,O13702,O14103/
9000  FORMAT (' ',2A6,I5,' CANNOT BE PROCESSED')
9001  FORMAT (' ',2A6,'MATRIX CANNOT BE READ')
9002  FORMAT (' ',10A6)
9003  FORMAT (' ','ERROR IN OBSERVATION',I5)
9004  FORMAT (' ','ACTUAL NUMBER OF OBSERVATIONS=',I7)
      IF (HNR.EQ.0) GO TO 100
         IF (HNR.LE.3) WRITE (6,9000) HDR(2*HNR-1),
     1   HDR(2*HNR),IND
         IF (HNR.GT.3) WRITE (6,9001) HDR(2*HNR-1),HDR(2*HNR)
100   IF (NR.GE.40) GO TO 110
         I1=FLD(21,9,SEEK(NR))
         I2=FLD(30,6,SEEK(NR))+I1-1
         WRITE (6,9002) (MSG(I),I=I1,I2)
         GO TO 200
110   I=NR-39
      GO TO (40,41),I
40    WRITE (6,9003) IND
      GO TO 200
41    WRITE (6,9004) IND
200   RETURN
      END
C 20
      SUBROUTINE SERINS(CODE,PLIST,NP,PROCED,PCODE)
C CALL SUBROUTINE PROCED ACCORDING TO PARAMETER LIST
      INCLUDE DEF,LIST
      INTEGER         CODE,NP,I,LOWER,UPPER,I1,PCODE,CODE1,
     1                STEP
      REAL            PLIST(NP)
      I=0
      LOWER=0
230   I=I+1
      IF (I.GT.NP) GO TO 240
         IF (PLIST(I).GE.0) LOWER=PLIST(I)
         UPPER=IABS(IFIX(PLIST(I)))
               STEP=1
               IF (LOWER.EQ.UPPER) GO TO 210
               IF (UPPER.LT.LOWER) STEP=-1
               LOWER=LOWER+STEP
210   DO 220 I1=LOWER,UPPER,STEP
         ,CALL PROCED(CODE,I1)
          IF (ERROR) GO TO 240
```

```
          IF (PCODE.EQ.0) GO TO 220
            CODE1=0
            IF (NP.GT.1) CODE1=CODE
            CALL OUTPUT (CODE1,I1,PCODE)
            IF (ERROR) GO TO 240
220   CONTINUE
      LOWER=UPPER
      GO TO 230
240   RETURN
      END
C 21
      SUBROUTINE HANVAR (CODE,IND1)
C ADD OR REMOVE A VARIABLE
      INCLUDE DEF,LIST
      INTEGER       CODE,IND,ADD,IA,ERNR,CHGIND
      ADD=1
      IF (CODE.EQ.2) ADD=-1
C CHECK THE INSTRUCTION
      IND=CHGIND (IND1)
      IF (IND.GT.0) GO TO 300
        ERNR=6
        GO TO 340
300   IF (ADD*K(IND).GT.0) GO TO 310
        ERNR=1
        GO TO 340
310   IF (DF-ADD.GE.0) GO TO 320
        ERNR=2
        GO TO 340
320   IF (ADD+INDEP-CONSTS.GE.0) GO TO 330
        ERNR=3
        GO TO 340
330   IF (.NOT.DVLIST(IND)) GO TO 345
        ERNR=4
        GO TO 340
C CHECK THE PIVOT ELEMENT
345   IA=R(IND)+IND
      IF (A(IA)) ,,360
        ERNR=5
C ERROR EXIT
340   ERROR=.TRUE.
      CALL  ERRMSG (ERNR,IND1,1)
      RETURN
C HANDLE THE VARIABLE
360   CONTINUE
```

```
350    CALL  PIVOT (IND)
       DF=DF-ADD
       INDEP=INDEP+ADD
       RETURN
       END
C 22
       SUBROUTINE  HANOBS (CODE,IND)
C ADD OR REMOVE AN OBSERVATION
       INCLUDE DEF,LIST
       INTEGER         CODE,IND,ADD,ERNR
       REAL            WEIGHT
       LOGICAL         ERR
       ADD=1
       IF (CODE.EQ.4) ADD=-1
C GET OBSERVATION AND CHECK INSTRUCTION
       CALL  SANDR (IND,WEIGHT,ERR)
       IF (.NOT.ERR) GO TO 400
           ERNR=7
           GO TO 420
400    IF (HOMOG) IF (DF+ISIGN(IFIX(WEIGHT),ADD))405,410,410
           IF (DF+ADD.GE.0) GO TO 410
405        ERNR=2
           GO TO 420
C CHECK WEIGHT
410    IF (WEIGHT.GE.0) GO TO 430
           ERNR=8
C ERROR EXIT
420    ERROR=.TRUE.
       CALL ERRMSG (ERNR,IND,2)
425    RETURN
C HANDLE THE OBSERVATION
430    IF (ADD.EQ.1.AND.DISC.NE.1) CALL DISCON (DISC)
435    CALL  GAUSS (ADD,WEIGHT)
       IF (.NOT.ERROR) GO TO 440
           ERNR=5
           GO TO 420
440    CALL  UPDATE (ADD,WEIGHT)
       LOBS=IND
       IF (ADD.EQ.1) GO TO 445
           LOBS=IND-1
           IF (DISC.NE.1) CALL  DISCON (1./DISC)
445    IF (HOMOG) GO TO 450
           OBS=OBS+ADD
           DF=DF+ADD
       RETURN
```

```
450   OBS=OBS+ADD*WEIGHT
      DF=DF+ADD*WEIGHT
      RETURN
      END

C 23
      SUBROUTINE HANCON (CODE,PLIST,NP)
C ADD OR REMOVE A CONSTRAINT
      INCLUDE        DEF,LIST
      INTEGER           CODE,IND,ADD,COLCON,ERNR,NP,IA
      REAL              WEIGHT,PLIST(NP)
      LOGICAL           ERR
      ADD=1
      IF (CODE.EQ.6) ADD=-1
      IND=PLIST(1)
C CHECK THE INSTRUCTION
      IF (DF+ADD.GE.0) GO TO 500
         ERNR=2
         GO TO 540
500   IF (INDEP.GE.CONSTS+ADD) GO TO 505
         ERNR=3
         GO TO 540
C CHECK IF THE CONSTRAINT WAS HANDLED BEFORE
505   IF (CONS.EQ.0) GO TO 525
      COLCON=ISEEK (IND)
      IF (COLCON.EQ.0) GO TO 525
         IF (ADD*K(COLCON).LT.0) GO TO 510
            ERNR=1
            GO TO 540
C CHECK THE PIVOT ELEMENT
510   IA=R(COLCON)+COLCON
      IF (A(IA)) ,,515
         ERNR=5
         GO TO 540
C ADD OR REMOVE THE CONSTRAINT HANDLED BEFORE
515   CALL  PIVOT (COLCON)
      GO TO 530
C ADD THE NEW CONSTRAINT
525   IF (ADD),,526
         ERROR=.TRUE.
         ERNR=9
         GO TO 540
526   IF (VARS+CONS.LT.P0-2) GO TO 520
         ERNR=10
         GO TO 540
```

```
520    CALL GETCON (ERR)
       IF ( ERR ) GO TO 545
           WEIGHT=1E+37
           CALL  GAUSS (ADD,WEIGHT)
           IF (.NOT.ERROR) GO TO 521
               ERNR=5
               GO TO 540
521        CONS=CONS+1
           NAME(VARS+CONS+1)=IND
C UPDATE COUNTERS
530    CONSTS=CONSTS+ADD
       DF=DF+ADD
       RETURN
C ERROR EXIT
540    CALL ERRMSG (ERNR,IND,3)
545    ERROR=.TRUE.
       RETURN
C ISEEK IS PUT EQUAL TO THE NUMBER OF THE ROW ATTACHED
C TO THE CONSTRAINT IF IT WAS HANDLED BEFORE,
C OTHERWISE THE VALUE IS ZERO
       FUNCTION  ISEEK(IND1)
       INTEGER          VARS2, VC1,I
       ISEEK=0
       VARS2=VARS+2
       VC1=VARS+CONS+1
       DO 550 I=VARS2,VC1
           IF (NAME(I).NE.IND1) GO TO 550
           ISEEK=I
           RETURN
550    CONTINUE
       RETURN
       SUBROUTINE GETCON (ERR)
C INTERPRET THE CONSTRAINT
       LOGICAL        ERR
       INTEGER      IX,I,CHGIND
       INTEGER        VARS1
       ERR=.FALSE.
       IF (NP.GT.2) GO TO 610
           ERR=.TRUE.
           CALL ERRMSG(9,IND,3)
           RETURN
610    VARS1=VARS+1
       DO 620 I=1, VARS1
620    X(I)=0.
       DO 600 I=2,NP,2
```

```
          IX= CHGIND (IFIX(PLIST(I)))
          IF  (IX.GT.0) GO TO 600
              ERR=.TRUE.
              CALL ERRMSG (6,IND,3)
              RETURN
600   X(IX)=PLIST (I+1)
      RETURN
      END
C 24
      SUBROUTINE  CHGWEI (PLIST,NP)
C CHANGE WEIGHT
      INCLUDE   DEF,LIST
      INTEGER          IND,DELTDF,NP,ERNR
      REAL             WEIGHT,DELTWG
      LOGICAL          ERR
      DIMENSION        PLIST(NP)
C TRANSFORM THE PARAMETER LIST
      IND=PLIST(1)
      DELTWG=PLIST(2)
      DELTDF=0
      IF (HOMOG)  DELTDF=IFIX(DELTWG)
      IF (NP.EQ.3) DELTDF=PLIST(3)
C GET OBSERVATION
      CALL  SANDR (IND,WEIGHT,ERR)
      IF (.NOT.ERR) GO TO 705
         ERNR=7
         GO TO 710
C CHECK THE DEGREERS OF FREEDOM
705      IF (DF+DELTDF.GE.0) GO TO 700
             ERNR=2
C ERROR EXIT
710   ERROR=.TRUE.
      CALL ERRMSG (ERNR,IND,2)
      RETURN
C CHECK WEIGHT
700   IF (ABS(DELTWG).GT.0.) GO TO 720
         ERROR=.TRUE.
         ERNR=8
         GO TO 710
720   CALL GAUSS (1,DELTWG)
      IF (.NOT.ERROR) GO TO 721
         ERNR=5
         GO TO 710
```

```
721    CALL UPDATE (1,DELTWG)
       DF=DF+DELTDF
       OBS=OBS+DELTDF
       RETURN
       END
C 25
       SUBROUTINE  PIVOT (H)
C PIVOTAL OPERATION
       INCLUDE DEF,LIST
       INTEGER          H,IJ,I,J,M1
       REAL             AHH,AHH1,AIH,AHJ
       DEFINE  AB(I)=A(I)
       DEFINE  AA(I,J)=AB(R(I)+J)
       M1=VARS+CONS+1
       AHH=AA(H,H)
5      AHH1=1./AHH
       DO  10  I=1,M1
           IF (I-H),10,20
               AIH=K(I)*K(H)*AA(H,I)
               GO TO 30
20         AIH=AA(I,H)
30         IJ=R(I)
           DO   40  J=1,I
                IJ=IJ+1
                IF (J-H),40,50
                    AHJ=AA(H,J)
                    GO TO 60
50              AHJ=K(H)*K(J)*AA(J,H)
60              A(IJ)=(A(IJ)*AHH-AIH*AHJ)*AHH1
40         CONTINUE
10     CONTINUE
C PIVOT ROW
       DO  70  J=1,M1
           IF  (J-H),70,90
               AA(H,J)=AA(H,J)*AHH1
               GO TO 70
90         AA(J,H)=-AA(J,H)*AHH1
70     CONTINUE
       AA(H,H)=AHH1
       K(H)=-K(H)
       RETURN
       END
C 26
       SUBROUTINE  GAUSS (ADD,WEIGHT)
C GAUSSIAN STEP
```

```
C IF WEIGHT IS 1E37 THEN HANDLE THE CONSTRAINT
      INCLUDE DEF,LIST
      INTEGER          IC,ADD,M1,M2,I,J
      REAL             G,DI,AIJ
      DEFINE     AB(I)=A(I)
      DEFINE     AA(I,J)=AB(R(I)+J)
      M1=VARS+CONS+1
      M2=VARS+1
      IF (WEIGHT.EQ.0) RETURN
      IC=R(M1+1)
      G=0.
      IF(WEIGHT.LT.1E37) G=1./(ADD*WEIGHT)
C CALCULATE VALUES OF DI
      DO 10 I=1,M1
         IC=IC+1
         DI=0.
         IF (K(I).GT.0.AND.I.LE.M2) DI=X(I)
         DO 20  J=1,M2
            IF (K(J)),,20
            IF (J.LE.I) AIJ=AA(I,J)
            IF (J.GT.I) AIJ=K(I)*K(J)*AA(J,I)
            DI=DI+AIJ*X(J)
20       CONTINUE
         IF (K(I).LT.0.AND.I.LE.M2) G=G+DI*X(I)
         A(IC)=-K(I)*DI
10    CONTINUE
C CHECK G AND COMPLETE THE STEP
      IC=IC+1
      A(IC)=G
      K(M1+1)=-1
      IF (G) 30,,30
         ERROR=.TRUE.
         RETURN
30    CALL PIVOT (M1+1)
      RETURN
      END

C 27
      SUBROUTINE  DISCON (DISCF)
C DISCOUNT THE REGRESSION MODEL
      INCLUDE DEF,LIST
      INTEGER    I,J,IJ,VC1
      REAL       DISCF1
      VC1=VARS+CONS+1
      IJ=1
```

```
1000  SW=DISCF*SW
      DISCF1=1./DISCF
      DO 1010 I=1,VC1
          SX(I)=DISCF*SX(I)
          SSX(I)=DISCF*SSX(I)
          DO 1020  J=1,I
              IF  (K(I).NE.K(J)) GO TO 1020
                  IF (K(I).EQ.-1) A(IJ)=DISCF1*A(IJ)
                  IF (K(I).EQ.1) A(IJ)=DISCF*A(IJ)
1020          IJ=IJ+1
1010  CONTINUE
      RETURN
      END

C 28
      SUBROUTINE UPDATE(ADD,WEIGHT)
C ACCUMULATE SUMS AND SUMS OF SQUARES
      INTEGER         ADD,VARS1
      INCLUDE DEF,LIST
      INTEGER     I
      REAL        WGH,WEIGHT
      WGH=ADD*WEIGHT
      SW=SW+WGH
      VARS1=VARS+1
      DO 1500 I=1,VARS1
      SX(I)=SX(I)+WGH*X(I)
      SSX(I)=SSX(I)+WGH*X(I)*X(I)
1500  CONTINUE
      RETURN
      END

C 29
      SUBROUTINE SANDR (IND,WEIGHT,ERR)
C SEEK OBSERVATION NO IND AND TRANSMIT IT TO X-VECTOR
      INCLUDE  DEF,LIST
      INTEGER  VARS1,IND,I1,I
      REAL     WEIGHT
      LOGICAL  ERR
C CHECK THE INDEX
      ERR=.FALSE.
      IF (IND.GE.1.AND.IND.LE.NOBS) GO TO 1900
         ERR=.TRUE.
         RETURN
1900  VARS1=VARS+NW
      I1=(IND-1)*VARS1
      DO 1910 I=1,VARS
```

```
1910 X(I+1)=OMATR(I1+I)
     WEIGHT=1.
     IF (WIND.NE.0.AND.NW.NE.0) WEIGHT=OMATR(I1+VARS1)
     X(1)=1.
     RETURN
     END
C 30
     SUBROUTINE CHGOUT(OCODE,PLIST,NP,ERROR)
C CHANGE PLIST TO OUTPUT CODE
     INTEGER        OCODE,NP,ERNR,I
     REAL           PLIST,CODE1
     LOGICAL        ERROR
     DIMENSION      PLIST(NP)
     OCODE=0.
     DO 900 I=1,NP
     CODE1=PLIST(I)
     IF (CODE1.NE.'N') GO TO 905
        OCODE=0.
        GO TO 900
905  IF (CODE1.NE.'VA') GO TO 910
        FLD(35,1,OCODE)=1
        GO TO 900
910  IF(CODE1.NE.'R') GO TO 920
        FLD(34,1,OCODE)=1
        GO TO 900
920  IF(CODE1.NE.'C') GO TO 930
        FLD(33,2,OCODE)=3
        GO TO 900
930  IF(CODE1.NE.'CT') GO TO 940
        FLD(32,3,OCODE)=7
        GO TO 900
940  IF(CODE1.NE.'F') GO TO 950
        FLD(31,1,OCODE)=1
        GO TO 900
950  IF(CODE1.NE.'RES') GO TO 960
        FLD(30,1,OCODE)=1
        GO TO 900
960  IF(CODE1.NE.'CORR') GO TO 970
        FLD(20,1,OCODE)=1
        GO TO 900
970  IF(CODE1.NE.'MOM') GO TO 975
        FLD(19,1,OCODE)=1
        GO TO 900
975  IF(CODE1.NE.'MS') GO TO 980
        FLD (18,1,OCODE)=1
```

```
900    CONTINUE
       RETURN
C UNDEFINED OUTPUT CLASS
980    ERROR=.TRUE.
       ERNR=12
       CALL ERRMSG (ERNR,CODE1,0)
       RETURN
       END
C 31
       SUBROUTINE OUTPUT (INSCOD,IND,OCODE)
C OUTPUT ACCORDING TO OCODE
       INCLUDE DEF,LIST
       PARAMETER      MAXHED=4
       INTEGER        INSCOD,IND,OCODE,VARS1,I
       REAL           HEADIN
       LOGICAL        SOMOUT
       DIMENSION      HEADIN (3,MAXHED)
       DATA HEADIN /
      1 'ADDITI','ON OF ','VAR   ',
      2 'REMOVA','L OF V','AR    ',
      3 'ADDITI','ON OF ','OBS   ',
      4 'REMOVA','L OF O','BS    '/
9005   FORMAT ('0',3A6,I5)
C HEADING
       IF (INSCOD.EQ.0.OR.INSCOD.GT.MAXHED ) GO TO 1000
          WRITE (6,9005) (HEADIN(I,INSCOD),I=1,3),IND
C MATRIX SPACE, CORREL MATRIX, MEANS AND STD DEVS
1000   IF (FLD(18,4,OCODE).EQ.0) GO TO 1010
          CALL OUTP1 (OCODE)
C R, REGR COEFFS, T- AND F-VALUES, RESIDUALS
1010   IF (FLD(30,6,OCODE).EQ.0) RETURN
          SOMOUT=.FALSE.
          VARS1=VARS+1
          DO 1020 I=1,VARS1
                IF (.NOT.DVLIST(I)) GO TO 1020
                    SOMOUT=.TRUE.
                    CALL OUTP2 (OCODE,I)
1020      CONTINUE
          IF (SOMOUT) RETURN
             CALL ERRMSG (11,0,0)
             RETURN
          END
C 32
       SUBROUTINE OUTP1 (OCODE)
C PRINT CORREL MATRIX, MATRIX SPACE, MEANS AND STD DEVS
```

```
      INCLUDE DEF,LIST
      PARAMETER    BS1=9,BS2=5
      INTEGER      HROW,VARS1,I
      REAL         ROW,S1,S2
      LOGICAL      SHORT
      DIMENSION    HROW(BS1),ROW(BS1)
9014  FORMAT ( '0MATRIX SPACE')
9015  FORMAT ('0CORRELATION MATRIX')
9024  FORMAT ('0NR OF OBSERVATIONS',I10//' VARIABLE',6X,'MEAN',
     16X,'STD DEV')
9025  FORMAT (' ',I5,3X,2G13.5)
      DEFINE  AB(I)=A(I)
      DEFINE  AA(I,J)=AB(R(I)+J)
C MEANS AND STANDARD DEVIATIONS
1200  IF (FLD(18,1,OCODE).EQ.0) GO TO 1210
         IF (OBS.LE.1) GO TO 1210
         WRITE (OUT,9024) OBS
         VARS1=VARS+1
         DO 1220 I=2,VARS1
               S1=SX(I)/SW
               S2=(SSX(I)-S1*SX(I))/(OBS-1)
               IF  (S2.GT.0) S2=SQRT(S2)
               IF  (S2.LT.0) S2=0.
1220     WRITE (OUT,9025) NAME(I),S1,S2
C CORRELATION MATRIX
1210  IF (FLD(20,1,OCODE).EQ.0) GO TO 1300
      IF (OBS.LE.1) GO TO 1300
      WRITE (OUT ,9015)
      SHORT=.TRUE.
      IF (K(1).EQ.-1) GO TO 1310
              SHORT=.FALSE.
              CALL PIVOT (1)
1310  CALL BLOUT (.TRUE.,BS1,2)
      IF (.NOT.SHORT) CALL PIVOT (1)
C MATRIX SPACE
1300  IF (FLD(19,1,OCODE).EQ.0) RETURN
         WRITE (OUT,9014)
         CALL BLOUT (.FALSE.,BS2,1)
         RETURN
C SUBROUTINE FOR PRINTING TRIANGULAR MATRIX
      SUBROUTINE  BLOUT (CORREL,BLSIZE,L1)
      INTEGER      BLSIZE,VARS1,LOWER,UPPER,U1,L1,IND,I,J
      REAL         A1
      LOGICAL      CORREL
```

```
9016  FORMAT ('0',3X,10I7)
9017  FORMAT ('0',I11,4I13)
9018  FORMAT (' ',I3,1X,10F7.3)
9019  FORMAT (' ',I3,5G13.5)
C HEADING
      LOWER=L1
      VARS1=VARS+1
      IF (.NOT.CORREL) VARS1=VARS1+CONS
1390  IND=0
      DO 1320 I=LOWER,VARS1
          UPPER=I
          IF(CORREL.AND.K(I).NE.1) GO TO 1320
          IND=IND+1
          HROW(IND)=NAME(I)
          IF (IND.GE.BLSIZE) GO TO 1330
1320  CONTINUE
1330  IF (IND.EQ.0) GO TO 1380
      IF (CORREL) WRITE(OUT ,9016) (HROW(I), I=1,IND)
      IF (.NOT.CORREL) WRITE(OUT ,9017) (HROW(I),I=1,IND)
C TRIANGULAR MATRIX
1340  DO 1350 I=LOWER,VARS1
          IF (CORREL.AND.K(I).NE.1) GO TO 1350
              IND=0
              U1=MIN0 (UPPER,I)
              DO 1360 J=LOWER,U1
                  IF (.NOT.CORREL) GO TO 1370
                      IF (K(J).NE.1) GO TO 1360
                          IND=IND+1
                          A1=AA(I,I)*AA(J,J)
                          ROW(IND)=0.
                          IF (A1.GT.0)ROW(IND)=AA(I,J)/SQRT(A1)
                          GO TO 1360
1370              IND=IND+1
                  ROW (IND)=AA(I,J)
1360          CONTINUE
              IF (CORREL) WRITE(OUT ,9018)NAME(I),(ROW(J1), J1=1
    1         ,IND)
              IF (.NOT.CORREL) WRITE(OUT ,9019)  NAME(I),(ROW(J1),
    1         J1=1,IND)
1350  CONTINUE
      LOWER=UPPER+1
      IF (UPPER.LT.VARS1) GO TO 1390
1380  RETURN
      END
```

```
C 33
      SUBROUTINE    OUTP2(OCODE,IND)
C PRINT R, REGR COEFF, T- AND F-VALUES, RESIDUALS
      INCLUDE  DEF,LIST
      INTEGER       OCODE,IND,NCOL,VARS1,DF1,DF2,I,J,I1,I2
      REAL          ROW,SE,SY,R1,SB,SF,SS1,SS2,SS3,SSE,SSR,SSY,
     1              RES
      LOGICAL       ERR
      DIMENSION     ROW(10)
9008  FORMAT ( '0R       =',G13.5/
     1        ' D.F.     =',I10)
9009  FORMAT ('0',8A6)
9010  FORMAT (' ',I5 ,3X,3G13.5)
9011  FORMAT ( '0VARIABLE',3X,'F-VALUE')
9012  FORMAT (' ',I5,A1,2X,G13.5)
9014  FORMAT ('0DEPENDENT VARIABLE',I5)
9015  FORMAT ('0ANALYSIS OF VARIANCE'//' VARIATION',10X,'SS',
     17X,'DF',7X,'MS',7X,'STD DEV'/
     2'   REGRESSION ',G13.5,I5,2G13.5/
     3'   RESIDUAL   ',G13.5,I5,2G13.5/
     4'   TOTAL      ',G13.5,I5,2G13.5/)
9016  FORMAT (' F VALUE      ',G13.5)
9017  FORMAT ('0RESIDUALS')
9018  FORMAT (I6,5G13.5)
9019  FORMAT ('0PREDICTIVE RESIDUALS'//'  NO   OBSERVED',
     1        5X,'PREDICTED',4X,'RESIDUAL')
      DEFINE  AB(I)=A(I)
      DEFINE  AA(I,J)=AB(R(I)+J)
      IF (DEPS.GT.1) WRITE (OUT,9014) NAME(IND)
      VARS1=VARS+1
C ANALYSIS OF VARIANCE AND R
      IF (FLD(34,2,OCODE).EQ.0.OR.OBS.LE.1) GO TO 1200
         DF1=OBS-DF-1
         DF2=OBS-1
         SS3=SSX(IND)-SX(IND)**2/SW
         SS2=AA(IND,IND)
         SS1=SS3-SS2
         SSE=0.
         IF(DF.GT.0.AND.SS2.GT.0) SSE=SS2/DF
         SSY=SS3/(OBS-1)
         R1=0.
         IF (SSY.GT.0.AND.SSE.LT.SSY) R1=SQRT(1.-SSE/SSY)
         IF (FLD(35,1,OCODE).EQ.0) GO TO 1205
         SSR=0.
```

```
      IF (DF1.GT.0.AND.SS1.GT.0) SSR=SS1/DF1
         IF (SSE.GT.0) FVA=SSR/SSE
         SE=SQRT(SSE)
         SY=SQRT(SSY)
         SR=SQRT(SSR)
         IF (FLD(34,1,OCODE).EQ.0) GO TO 1200
1205     WRITE (OUT,9008) R1,DF
C REGR COEFF, T-VALUES
1200  IF (FLD(32,2,OCODE).EQ.0) GO TO 1210
         NCOL=4
         IF(FLD(32,1,OCODE).EQ.1) NCOL=8
C HEADING LINE
         ROW(1)='VARIAB'
         ROW(2)='LE     '
         ROW(3)=' COEFF'
         ROW(4)='       '
         ROW(5)='STD  DE'
         ROW(6)='V      '
         ROW(7)='  T-VAL'
         ROW(8)='UE     '
         WRITE (OUT ,9009) (ROW(I),I=1,NCOL)
         NCOL=NCOL/2-1
C REGRESSION COEFFICIENTS
      DO 1230 I=1,VARS1
         IF (K(I)),,1230
            IF (I.GT.IND) ROW(1)=AA(I,IND)
            IF (I.LT.IND) (ROW(1)=K(IND)*K(I)*AA(IND,I)
1250     IF (NCOL.LE.1) GO TO 1260
C STANDARD DEVIATIONS
         SB=AA(I,I)
         IF (SB.LT.0) SB=0.
1270     ROW(2)=SQRT(SB*SSE)
C T-VALUES
         ROW(3)=0.
         IF ( ROW(2).NE.0)ROW(3)=ROW(1)/ROW(2)
1260     WRITE(OUT ,9010) NAME(1),(ROW(J),J=1,NCOL)
1230  CONTINUE
C F-VALUES
1210  IF (FLD(31,1,OCODE).EQ.0) GO TO 1295
         WRITE (OUT ,9011)
         DO 1280 I=1,VARS1
             IF (I.EQ.IND) GO TO 1280
1290         ROW(1)=''
             IF (K(I).LE.0) ROW(1)='*'
```

```
              ROW(2)=FVALUE(IND,I)
              WRITE (OUT ,9012) NAME(I), ROW(1),ROW(2)
1280      CONTINUE
C THE ANALYSIS OF VARIANCE
1295  IF (FLD(35,1,OCODE).EQ.0) GO TO 1285
          WRITE (OUT,9015) SS1,DF1,SSR,SR,SS2,DF,SSE,SE,
     1                      SS3,DF2,SSY,SY
          IF (K(1).GT.0.AND.A(1).GT.1E-6) GO TO 1284
              WRITE (OUT,9016) FVA
              GO TO 1285
1284      CALL ERRMSG (13,0,0)
1285  CONTINUE
C RESIDUALS
      IF (FLD(30,1,OCODE).EQ.0) RETURN
      IF (NPR.GT.0) GO TO 1330
          WRITE (OUT,9017)
          DO 1300 I=1,NOBS,5
                I1=MIN0(NOBS,I+4)
                DO 1310 J=I,I1
                    CALL SANDR (J,WEIGHT,ERR)
                    CALL YESTIM (YEST)
1310              ROW(J-I+1)=X(IND)-YEST
                  I2=I1-I+1
1300      WRITE (OUT,9018)I,(ROW(J),J=1,I2)
      RETURN
C PREDICTIVE RESIDUALS
1330  WRITE (OUT,9019)
      I1=LOBS+1
      I2=LOBS+NPR
      DO 1340 I=I1,I2
          CALL SANDR (I,WEIGHT,ERR)
          IF (ERR) GO TO 1350
              CALL YESTIM (YEST)
              RES=X(IND)-YEST
1340          WRITE (OUT,9018)I,X(IND),YEST,RES
1350  RETURN
      SUBROUTINE YESTIM (YEST)
C CALCULATE ESTIMATE OF Y
      INTEGER J,VARS1
      VARS1=VARS+1
      YEST=0.
      DO 1360 J=1,VARS1
          IF (K(J).GT.0) GO TO 1360
          IF (J.GT.IND) YEST=YEST+AA(J,IND)*X(J)
          IF (J.LT.IND) YEST=YEST-AA(IND,J)*X(J)
```

```
1360  CONTINUE
      RETURN
      END
C 34
      REAL FUNCTION FVALUE (IY,IX)
C COMPUTE F-VALUES
C IY IS THE INDEX OF THE DEPENDENT VARIABLE
C IX IS THE INDEX OF THE INDEPENDENT VARIABLE
      INCLUDE DEF,LIST
      DEFINE    AB(I)=A(I)
      DEFINE    AA(I,J)=AB(R(I)+J)
      REAL      D,DELTA
      INTEGER   IX,IY
      IF (AA(IX,IX).GT.0) GO TO 1400
         FVALUE=0
         RETURN
1400     IF (IX.GT.IY) DELTA= AA(IX,IY)**2/AA(IX,IX)
         IF (IY.GT.IX) DELTA= AA(IY,IX)**2/AA(IX,IX)
         IF (K(IX).EQ.1) GO TO 1410
         FVALUE=(DF*DELTA)/AA(IY,IY)
         RETURN
1410  D=(AA(IY,IY)-DELTA)
      IF (D.NE.0) GO TO 1420
         FVALUE=1.E+37
         RETURN
1420  FVALUE=(DF-1)*DELTA/D
      RETURN
      END
C 40
      SUBROUTINE OPENL (LIST,NP,OLIST,I1,NP1,ERR,ULIMIT)
C INTERPRET THE INDEX LIST
      INTEGER  ULIMIT,OLIST(ULIMIT),NP,I1,NP1,UPPER,LOWER,I,K
      LOGICAL  ERR
      REAL     LIST(NP)
      NP1=0
      I=0
      LOWER=1
230   I=I+1
      IF (I.GT.NP) GO TO 240
         IF (LIST(I).GT.0) LOWER=LIST(I)
            UPPER=IABS(IFIX(LIST(I)))
                  IF(UPPER.GE.LOWER) GO TO 210
                       ERR=.TRUE.
                       RETURN
```

```
210                 DO 220 K=LOWER,UPPER
                        IF (I1+NP1.LE.ULIMIT) GO TO 215
                           ERR=.TRUE.
                           RETURN
215                     OLIST (I1+NP1)=K
220                     NP1=NP1+1
           LOWER=UPPER+1
           GO TO 230
240     RETURN
        END

C 41
        SUBROUTINE CHGPL (NAME, NAMES,PLIST,M1,M2,ERR)
C CONSTRUCT PICKLIST PLIST FROM LISTS NAME AND NAMES
        INTEGER      M1,M2,NAME(M1),NAMES(M2),PLIST(M1),I
        LOGICAL      ERR
        PLIST(1)=ISEEK1(NAME(2),0,Å1800)
        DO 1810 I=2,M1
1810    PLIST(I)=ISEEK1(NAME(I+1),PLIST(I-1),Å1800)
1800    RETURN
        FUNCTION ISEEK1 (IND,LB,Å)
C SEEK VARIABLE IND FROM LIST NAMES
        INTEGER         I1,I2,LB,IND
        I1=LB+1
        IF (I1.GT.M2) GO TO 1850
        I2=M2
1840    DO 1820 I=I1,I2
        IF (NAMES(I).NE.IND) GO TO 1820
           ISEEK1=I
           RETURN
1820    CONTINUE
        IF (I1.EQ.1) GO TO 1830
1850       I2=I1-1
           I1=1
           GO TO 1840
C VARIABLE NOT FOUND
1830    ERR=.TRUE.
        RETURN 3
        END
```

6. A Program for Data Transformation

6.1. Description of the program

GENERAL

The program has been written for the UNIVAC 1108 computer using FORTRAN V language. The name of the program is MTL*F.TR. The program is both time sharing and batch processing oriented. In this description, the terminology attached to time sharing is adopted.

FUNCTION

The program prepares input material for the program MTL*F.RA (or MTL*F.RB) for stepwise regression analysis. An observation matrix is subjected to transformations, and a transformed observation matrix and/or the corresponding moment matrix, or central moment matrix, is produced.

INPUT

As input material, an observation file similar to that in the program MTL*F.RA is used.

TRANSFORMATIONS

The program consists of two parts, the ready coded main program MTL*F.TR and the subprogram TRANSF which is to be coded by the user. The main program reads the observation matrix by rows, and after reading any row it calls the subprogram, which performs the desired transformations. The main program accumulates the moment matrix, and generates the output file(s).

The subprogram to be coded by the user is of the following form (the cards are punched from column 7 on):

```
SUBROUTINE TRANSF(X,K)
DIMENSION X(200)
⟨instructions⟩
END
```

The elements x_{ij}, $j = 1, \dots, m$, of observation i are read into the vector $X(200)$ as follows

$$X(1) = x_{i1}, X(2) = x_{i2}, \dots, X(m) = x_{im} \,.$$

⟨instructions⟩ consists of ordinary FORTRAN statements, which accomplish the desired transformations to $X(1) \cdots X(m)$. The results are to be stored in $X(1) \cdots X(200)$. The transformed variables may again be used as operands in further transformations. In addition to transforming instructions there may, of course, be other FORTRAN statements: unconditional and conditional jumps, IF-statements etc. Sometimes, ⟨instructions⟩ is not needed, for example if the moment matrix of the original variables is to be calculated. The acceptance and rejection of observations is controlled by the parameter K, which may be given the following values:

K	**Meaning**
0	The observation is accepted in all output files
1	The observation is rejected from all output files
2	The observation is rejected from the transformed observation file
3	The observation is rejected from the moment file

The assumed value of K, attached to any observation, is 0, i.e. if K is not given, the program gives it a value of zero. An example of the subprogram is given in ILLUSTRATION.

OPERATING

Definition of files. The input file is defined as follows

```
#ASG␣⟨input file name⟩
#USE␣20,⟨input file name⟩
```

Correspondingly, the output files are defined as follows

```
#ASG␣⟨observation file name⟩
#USE␣10,⟨observation file name⟩
#ASG␣⟨moment file name⟩
#USE␣11,⟨moment file name⟩
```

Thus the original observation file is located on file no 20, and the transformed observation file and the moment file (or central moment file), if any, on files no 10 and 11. These definitions are compatible with those attached to the program MTL∗F.RA (or MTL∗F.RB).

Execution. Before execution of the program, the subprogram TRANSF must be compiled and linked with the main program. An example of the procedure is as follows:

(i) The subprogram, coded by the user, is compiled and located as element TRANSF in the provisional file TPF$. If the program is written on paper tape or on punch cards, this is accomplished as follows:

```
#FOR,I␣,.TRANSF
```
⟨the subprogram⟩

If, however, the subprogram (in FORTRAN) is located, say, as element ELEM in file PROGFILE (defined by the user), this phase is carried through by

```
#FOR␣PROGFILE.ELEM,TPF$.TRANSF.
```

(ii) The subprogram is linked with the main program by giving

```
#MAP␣,.TR
IN␣MTL*F.TR
IN␣.TRANSF
```

(iii) The execution instruction

```
#XQT␣.TR
```

is given. Then the program asks OUTPUT SPECIFICATION. As the answer, an output specification, introduced below, is given.

Output specification. The output is controlled by a specification card, of the form of a FORTRAN name list (cf. ⟨header⟩ in MTL*F.RA):

␣&TR␣PICK=⟨list⟩,
TM=s_1,
MM=s_2,
CM=s_3,
W=k,
CL=t,
RENAME=⟨list⟩␣&END

⟨list⟩ is of the same form as in MTL*F.RA. In the first line, the elements of ⟨list⟩ specify the elements of X (i.e. certain variables), which are to be included in output. The assumption of PICK=⟨list⟩ is PICK=1,$-m$.

TM, MM and CM are short for transformed observation matrix, moment matrix and central moment matrix. Any matrix is or is not generated according to whether the corresponding s_i equals 1 or 0. However, only one of the values s_2 and s_3 may be 1. The assumed value of each s_i is 0.

The specifiers W and CL are needed only in the case of weighting. k (≤ 200) indicates that the weight of any observation is given by the $X(k)$ attached to it. If k does not appear in the ⟨list⟩ on the first line of the output specification above, the variable $X(k)$ is automatically taken as the last variable into the transformed observation file. t is given a value of one or zero according to whether the material is classified or nonclassified, see the description of MTL*F.RA. The assumption of CL$=t$ is CL$=1$.

The specifier RENAME is given if it is desired to renumber the variables. The indices of the ⟨list⟩ on the last line are attached, in order, to the variables mentioned in the ⟨list⟩ on the first line.

It should be noted that the last specifier (before &END) is followed by a blank, the comma not being necessary.

LIMITATIONS

The program requires that

$$m, m_1 \leqq 200\,,$$
$$m_2 \leqq 50\,,$$

where m, m_1 and m_2 are the numbers of the variables in the input file, in the transformed observation file and in the moment file respectively. It should be noted that the possible weighting variable is included in the transformed observation file.

MESSAGES

After a normal course the program gives the messages

XX OBSERVATIONS INCLUDED IN INPUT FILE
XX OBSERVATIONS ACCEPTED TO OBSERVATION FILE
XX OBSERVATIONS ACCEPTED TO MOMENT FILE

Of course, the second or third message may be omitted. In the case of an abnormal course, one of the following error messages is given by the program:

**ERROR IN OUTPUT SPECIFICATION
EXECUTION TERMINATED
**ERROR IN THE HEADER OR FORMAT SPECIFICATION
OF THE INPUT FILE
EXECUTION TERMINATED

```
**ERROR IN OBSERVATION XX
   EXECUTION TERMINATED
**NEGATIVE WEIGHT
   OBSERVATION XX REJECTED
```

For the second error message, see the description of the program MTL*F.RA, pp. 30 and 41–2.

ILLUSTRATION

Reference is made to the numerical example given in the description of the program MTL*F.RA. Suppose that the observation file FDATA has been generated from punch cards in a batch run. The variables are subjected to the following transformations:

$$x_4 = \sqrt{x_1^2 + x_2^2}\,,$$

$$x_5 = \log x_4\,,$$

$$x_6 = \log x_3\,.$$

Observations are to be rejected from the transformed observation file if $x_3 \leqq 1$, and from the moment file if $x_3 > 1$. As output, the transformed observation file and the moment file are desired, including the variables x_3, x_4, x_5 and x_6 (in this order). The execution of the program is as follows:

```
#RUN RUN1,ACCOUNT1,PROJ1
DATE: 102472          TIME: 131627
#PSW

PSW=88888888888
READY
#ASG,A FDATA
READY
#ASG,T FTD,F2
READY
#ASG,T FMM,F2
READY
#USE 20,FDATA.
READY
#USE 10,FTD
READY
#USE 11,FMM
READY
#FOR,I ,.TRANSF
FOR 010A—10/24/72—13:18:30
```

```
**TAPE START**
        SUBROUTINE TRANSF(X,K)
        DIMENSION X(200)
        X(4)=SQRT(X(1)**2+X(2)**2)
        X(5)=ALOG10(X(4))
        X(6)=ALOG10(X(3))
        IF(X(3).LE.1)K=2
        IF(X(3).GT.1)K=3
        RETURN
        END

END OF TAPE
#


   STORAGE USED: CODE(1) 000063; DATA(0) 000007; BLANK
                 COMMON(2) 000000

END FOR
#MAP,N ,.TR
MAP 0023-10/24-13:19

IN MTL*F.TR
IN .TRANSF

#XQT .TR

OUTPUT SPECIFICATION
&TR PICK=3,-6,TM=1,MM=1  &END

      10  OBSERVATIONS INCLUDED IN INPUT FILE
       3  OBSERVATIONS ACCEPTED TO OBSERVATION FILE
       7  OBSERVATIONS ACCEPTED TO MOMENT FILE
     .
     .
     .
#FIN


    RUNID: RUN1     ACCOUNT: ACCOUNT1     PROJECT: PROJ1
TIME: 00:00:02.022           IN: 100     OUT: 0          PAGES:   1
    INITIATION TIME:    13:16:27-OCT 24,1972
    TERMINATION TIME:   13:23:42-OCT 24, 1972

                                  ****LINE INACTIVE****
```

The dots at the end of the above listing indicate that the program MTL*F.RA is carried through immediately after the transformation program. It is, of course, also possible to save the output file(s) of the transformation program, and to carry through the program MTL*F.RA later on. (Then, in the definitions #ASG,T␣FTD,F2 and #ASG, T␣FMM,F2 at the beginning of the program, the specifier T must be replaced by C.)

6.2. Technical notes

1° Construction of the moment matrix. The moment matrix is of the form

⟨header⟩
⟨sum of weights⟩
⟨sums of variables⟩
⟨sums of squares of variables⟩
⟨moments⟩

⟨header⟩ is of the same form as in the observation file (p. 30), augmented by the specifier FID which is put equal to 2 or 3 according to whether the matrix consists of moments or central moments. The moments are saved in vector A in the form of a lower triangular matrix.

2° Accumulation of central moment matrix. For accumulation of the central moment matrix, the auxiliary variables A_{0j}, $j = 0, \ldots, m$, are adopted, where m is the number of variables. The first observation is accumulated by the formulae

$$\begin{aligned} A_{00} &= 1/w_1 , \\ A_{0j} &= -z_{1j} , \qquad 1 \leqq j \leqq m , \\ A_{ij} &= 0 , \qquad 1 \leqq j \leqq i \leqq m . \end{aligned}$$

Observation no λ ($\lambda > 1$) is accumulated by the formulae

$$\begin{aligned} g &= A_{00} + 1/w_\lambda , \\ A_{00}^+ &= A_{00} - A_{00}^2/g , \\ A_{0j}^+ &= A_{0j} - (z_{\lambda j} + A_{0j})\, A_{00}/g , \qquad 1 \leqq j \leqq m , \\ A_{ij}^+ &= A_{ij} + (z_{\lambda i} + A_{0i})\,(z_{\lambda j} + A_{0j})/g , \qquad 1 \leqq j \leqq i \leqq m . \end{aligned}$$

3° A comment on weighting. If both a transformed observation matrix and a weighted moment matrix are desired, and the index of the weighting variable is not included in the list PICK, then the weighting variable is placed into the transformed observation matrix as the last variable.

6.3. Program listing

```
C 50
C PROGRAM FOR DATA TRANSFORMATION AND/OR FOR ACCUMU-
C LATING (CENTRALIZED) MOMENT MATRIX
      PARAMETER PAR1=200 , PAR2=50,PAR3=(PAR2+1)*PAR2/2,
     1              OUT=6
      REAL            PICK(PAR1),RENAME(PAR1),FORM(72),
     1                X(PAR1), XX(PAR1)
      INTEGER         NUMER/1/,TM,MM,CM,PLIST(PAR1),UPPER,
     1                FID,NF,
     2                REJECT,W,CL,NAMES(PAR1),NP2,NP3,NP4
      LOGICAL         ERR
      NAMELIST        /TR/PICK,TM,MM,CM,W,CL,RENAME
      NAMELIST        /HDR/FID,M,N,NAMES,NF,CL
      PARAMETER       F10=10,F11=11,F20=20,F5=5
      REAL            SX(PAR2),SSX(PAR2),A0(PAR2),A(PAR3)
9000  FORMAT ('0OUTPUT SPECIFICATION')
9002  FORMAT (13A6,A2)
9003  FORMAT ('0**ERROR IN OUTPUT SPECIFICATION'/
     1'     EXECUTION TERMINATED')
9005  FORMAT (12A6)
9006  FORMAT ('0',I7,' OBSERVATIONS INCLUDED IN INPUT FILE')
9007  FORMAT ('(10A6)')
9008  FORMAT ('0**ERROR IN THE HEADER OR FORMAT SPECIFI'
     1,'CATION',
     2' OF THE INPUT FILE'/' EXECUTION TERMINATED')
9009  FORMAT ('0**ERROR  IN OBSERVATION',17/' EXECUTION',
     1'TERMI NATED')
9010  FORMAT ('0**NEGATIVE WEIGHT'/' OBSERVATION',I7,
     1' REJECTED')
9011  FORMAT ('0',I7,' OBSERVATIONS ACCEPTED TO OBSERVATION'
     1,' FILE')
9012  FORMAT ('0',I7,' OBSERVATIONS ACCEPTED TO MOMENT',
     1'FILE')
      CL=1
      NF=1
      FID=1
C INPUT PARAMETERS
      WRITE (OUT,9000)
      READ (F5,TR,END=520,ERR=520)
      IF (CL.NE.0.AND.CL.NE.1) GO TO 520
      NP=0
```

```
      DO 110 I=1,PAR1
          IF (PICK(I).EQ.0) GO TO 104
110   NP=NP+1
104   DO 111 I=1,PAR1
          IF (RENAME(I).EQ.0) GO TO 105
111       NP2=NP2+1
C READ THE HEADER OF INPUT FILE AND CHECK PARAMETERS
105   READ (F20,HDR,ERR=500,END=500)
      IF (FID.NE.1.OR.NF.GT.6) GO TO 500
      IF (W.LT.0.OR.W.GT.PAR1) GO TO 520
C READ FORMAT CARDS
150   UPPER=NF*12
      READ (F20,9002,ERR=500,END=500)(FORM(I),I=1,UPPER)
C IF PICK HAS NOT BEEN DEFINED ASSUME PICK=1,-M
      IF (NP.EQ.0) GO TO 155
         CALL OPENL (PICK,NP,PLIST,1,NP1,ERR,PAR1)
         IF (ERR) GO TO 520
         GO TO 175
155      DO 160 I=1,M
160      PLIST(I)=I
         NP1=M
175   IF (CM+MM.NE.0.AND.NP1.GT.PAR2) GO TO 520
C SHOULD WEIGHTING VARIABLE BE PICKED
      NP4=NP1
      IF (W.EQ.0) GO TO 185
         DO 190 I=1,NP1
             IF (W.EQ.PLIST(I)) GO TO 185
190      CONTINUE
         PLIST(NP1+1)=W
         NP4=NP1+1
C CONSTRUCT NAMES VECTOR
185   IF (NP2.NE.0) GO TO 170
         DO 165 I=1,NP4
165      NAMES(I)=PLIST(I)
         GO TO 180
170   CALL OPENL(RENAME,NP2,NAMES,1,NP3,ERR,PAR2)
         IF (ERR.OR.NP3.NE.NP4) GO TO 520
C CHECK OUTPUT MODE
180   IF (MM*CM.NE.0) GO TO 520
C WRITE HEADER
      IF(TM.EQ.0) GO TO 215
      CALL HEADER (F10,1,NP4)
      WRITE (F10,9007)
C HANDLE THE OBSERVATION MATRIX
```

```
215   NOBS=0
      NACC=0
      NACC1=0
      MSIZE=NP1*(NP1+1)/2
      DO 220 I=1,N
           READ (F20,FORM,ERR=510,END=240)(X(J),J=1,M)
           NOBS=NOBS+1
      REJECT=0
      CALL TRANSF (X,REJECT)
      IF (REJECT.EQ.1) GO TO 220
          WEIGHT=1.
          IF (W.EQ.0) GO TO 225
              WEIGHT=X(W)
                IF (WEIGHT.GE.0) GO TO 225
                   WRITE (OUT,9010) I
                   GO TO 220
225       DO 230 J=1,NP4
          J1=PLIST(J)
230       XX(J)=X(J1)
          IF (TM.EQ.0.OR.REJECT.EQ.2) GO TO 531
              NACC=NACC+1
              CALL OUTPUT (F10,XX,NP4)
531           IF (MM.EQ.0) GO TO 532
              IF (REJECT.EQ.3) GO TO 220
               NACC1=NACC1+1
               CALL ACCUM (1)
               GO TO 220
532       IF (CM.EQ.0.OR.REJECT.EQ.3) GO TO 220
              NACC1=NACC1+1
              IF (WEIGHT.NE.O) CALL ACCUM (2)
220   CONTINUE
C WRITE MOMENT FILE
240   IF (MM+CM.EQ.0) GO TO 250
          N=NACC1
          IF (MM.NE.0) CALL HEADER (F11,2,NP1)
          IF (CM.NE.0) CALL HEADER (F11,3,NP1)
          WRITE (F11,9005) SW
          CALL OUTPUT (F11,SX,NP1)
          CALL OUTPUT (F11,SSX,NP1)
          CALL OUTPUT (F11,A,MSIZE)
C END MESSAGE
250   WRITE (OUT,9006) NOBS
      IF (TM.NE.0) WRITE(OUT,9011) NACC
      IF (CM+MM.NE.0) WRITE(OUT,9012) NACC1
      STOP
```

```
C ERROR EXIT
500   WRITE (OUT,9008)
      STOP
510   WRITE (OUT,9009) NOBS
      STOP
520   WRITE (OUT,9003)
      STOP
      SUBROUTINE  ACCUM (IT)
C ACCUMULATE SUMS, SUMS OF SQUARES  AND, IF IT=1, THE
C MOMENT MATRIX, OTHERWISE (IT=2) THE CENTRALIZED
C MOMENT MATRIX
      LOGICAL       FCALL/.TRUE./
      INTEGER       IT,I,J,IJ
      DO 260  I=1,NP1
         SX(I)=SX(I)+XX(I)*WEIGHT
260      SSX(I)=SSX(I)+XX(I)*XX(I)*WEIGHT
         SW=SW+WEIGHT
      IF (IT.NE.1) GO TO 270
         IJ=1
         DO 280 I=1,NP1
         DO 280 J=1,I
         A(IJ)=A(IJ)+XX(I)*XX(J)*WEIGHT
280      IJ=IJ+1
      RETURN
C CENTRALIZED MOMENTS
270   IF (.NOT.FCALL) GO TO 290
         FCALL=.FALSE.
         A00=1./WEIGHT
         DO 300 I=1,NP1
300      A0(I)=-XX(I)
      RETURN
290   G1=1./(1./WEIGHT+A00)
      G2=G1*A00
      IJ=1
      DO 310 I=1,NP1
          XX (I)=XX(I)+A0(I)
          A0(I)=A0(I)-XX(I)*G2
          DO 310 J=1,I
                A(IJ)=A(IJ)+XX(I)*XX(J)*G1
310             IJ=IJ+1
      A00=A00-A00*G2
      RETURN
      SUBROUTINE  OUTPUT (F,X,NX)
```

```
C WRITE VECTOR X IN BLOCKED FORM
      PARAMETER  BLC=10
      INTEGER  NX,I,I1,I2,F
      REAL  X(NX)
9005  FORMAT (10A6)
      DO 320 I1=1,NX,BLC
          I2=MIN0 (NX,I1+BLC-1)
320   WRITE (F,9005) (X(I),I=I1,I2)
      RETURN
      SUBROUTINE  HEADER (F,ID,NNP)
      INTEGER  F,ID
9010  FORMAT (' &HDR FID=',I5,',M=',I5,',N='1',
     1',CL=',I5,',NAMES=')
9011  FORMAT (20(10(I5,',')/))
9012  FORMAT ( '      &END')
      WRITE (F,9010) ID,NNP,N,CL
      WRITE (F,9011) (NAMES(I),I=1,NNP)
      WRITE (F,9012)
      RETURN
      SUBROUTINE OPENL (LIST,NP,OLIST,I1,NP1,ERR,ULIMIT)
C INTERPRET THE INDEX
      PARAMETER P0=50
      INTEGER   OLIST(P0),NP,I1,NP1,UPPER,LOWER,I,K,ULIMIT
      LOGICAL   ERR
      REAL      LIST(NP)
      NP1=0
      I=0
      LOWER=1
230   I=I+1
      IF (I.GT.NP) GO TO 240
         IF (LIST(I).GT.0) LOWER=LIST(I)
             UPPER=IABS(IFIX(LIST(I)))
                   IF(UPPER.GE.LOWER) GO TO 210
                        ERR=.TRUE.
                        RETURN
210                DO 220 K=LOWER,UPPER
                        IF (NP1.LT.ULIMIT) GO TO 215
                           ERR=.TRUE.
                           RETURN
215                     OLIST (I1+NP1)=K
220                     NP1=NP1+1
             LOWER=UPPER+1
             GO TO 230
240   RETURN
      END
```

References

1. DRAPER, N., and SMITH, H.: Applied Regression Analysis, John Wiley & Sons, Inc., New York, London and Sydney, 1966.
2. GOLDBERGER, A. S.: Econometric Theory, John Wiley & Sons, Inc., New York, London and Sydney, 1964.
3. TÖRNQVIST, L.: A Method for Calculating Changes in Regression Coefficients and Inverse Matrices Corresponding to Changes in the Set of Available Data, Skandinavisk Aktuarietidskrift (1957) 219–26.
4. VÄLIAHO, H.: A Synthetic Approach to Stepwise Regression Analysis, Comment. Phys.-Math. **34** (1969) 91–132.
5. VÄLIAHO, H.: Non-optimal Weighting in Stepwise Regression Analysis, Comment. Phys.-Math. **41** (1971) 63–72.
6. VÄLIAHO, H.: New Algorithms for Quadratic Programming, Comment. Phys.-Math. **36** (1970) 111–36.
7. WENDLER, K.: Hauptaustauschschritte (Principal Pivoting), Lecture Notes in Operations Research and Mathematical Systems, Springer-Verlag, Berlin, Heidelberg, New York, 1971.

Bei Fragen zur Produktsicherheit wenden Sie sich bitte an:
If you have any questions regarding product safety,
please contact:

Walter de Gruyter GmbH
Genthiner Straße 13
10785 Berlin
productsafety@degruyterbrill.com